Ethiopian Episodes: Two years, Two suitcases

Two ordinary people in one extraordinary place

Simon Noble

Ethiopian Episodes: Two years, Two suitcases

ISBN: 978-1-7645257-0-1

ደስተኛ ሁን ማለት ሁሉ ይኑርህ ማለት አይደለም፤ ባለህ ነገር ደስ ይበልህ ማለት እንጂ!

"Being happy doesn’t mean having everything; it means being content with what you have."

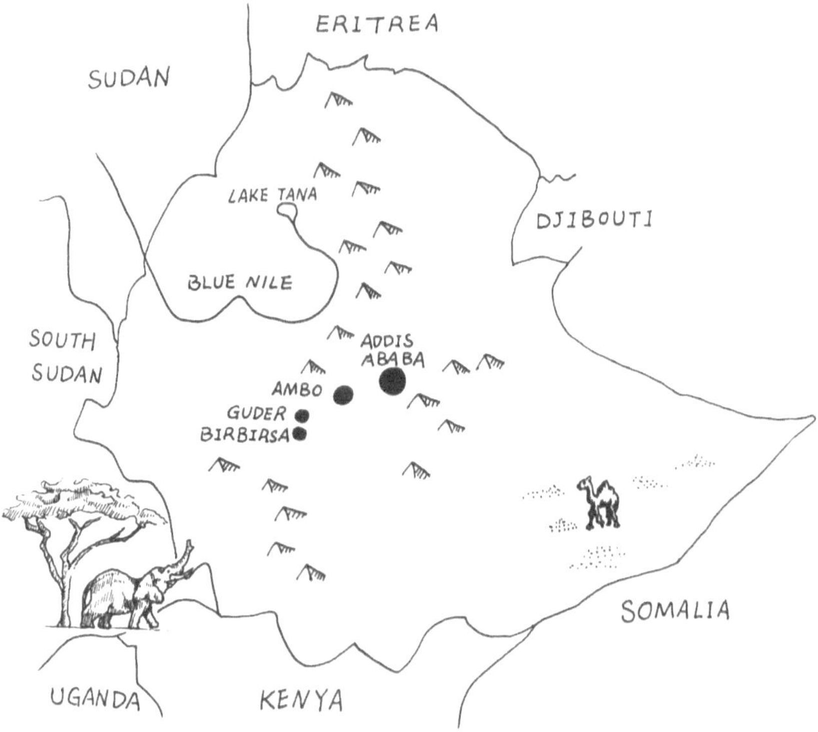

Ethiopia with the Farm Location at Birbirsa

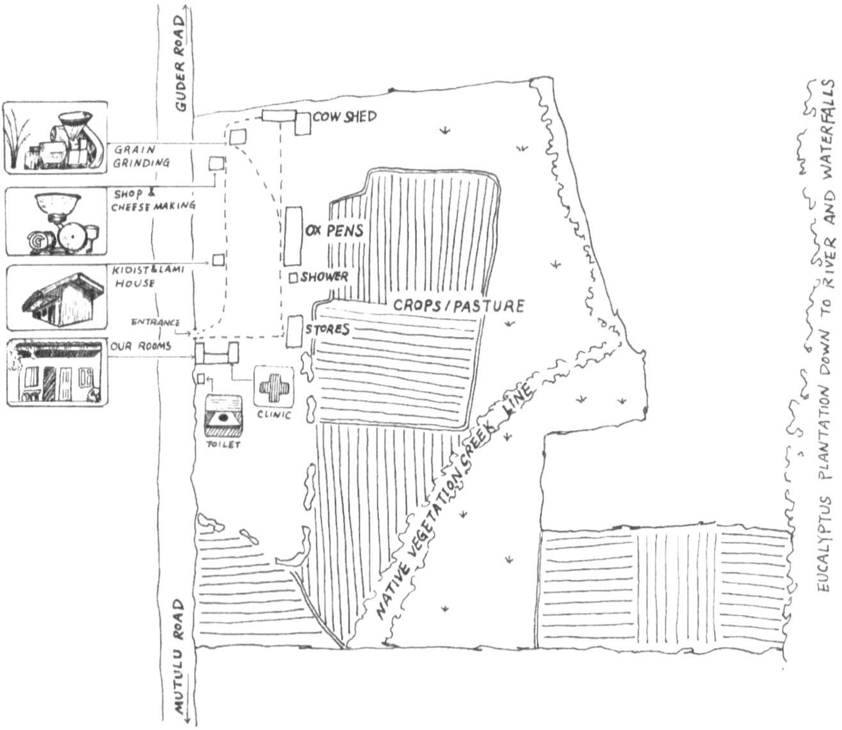

Layout of the Farm at Birbirsa

Chapter 1

The Dilemma

When the heart says yes and the head screams no— ***absolutely NO*** *—common sense insists the head should win. Yet some decisions refuse to be settled by logic alone.*

In 2013, my wife Pip and I faced that dilemma and chose the heart. It carried us far from our farm in northeast Victoria, Australia, to a remote area of western Ethiopia, where we would live and work for the next two years.

Ethiopia met us without compromise. Each day led to experiences we could never have imagined: moments of love, exhilaration and joy, stretches of frustration and despair, and encounters with hunger, suffering, life and death. With every passing week, our certainties were challenged and our moral compass recalibrated.

Almost by accident, we began writing from the first day. Remarkably, we were still writing seven hundred and thirty days later. These daily reflections—what we came to call our Ethiopia Episodes—became a record of our lives - lived at the edge of familiarity. Though it is impossible to capture every story, these pages reflect our journey from being Australian farmers to becoming international aid workers, mentoring farm managers in one of the most remote agricultural regions in Africa.

This is the story of that journey—and of a country that changed us in ways we could never have anticipated

By pure chance, an email caught the eye of Pip which mentioned volunteer roles in Africa. Could that be a good fit for her, a retired agricultural scientist and for husband, farmer Simon?

Following the link to the website of Australian Volunteers International (AVI) it became obvious that details for the role where very sketchy and the living arrangements even more so. 'Farm management mentors' the document read, but what of the accommodation? An unfurnished hut, a hundred and fifty kilometres from Addis Ababa the capital city of Ethiopia and, to ensure a favourable flood of inquiry, the promise of a squat latrine and no hot water!!

As we sat and chatted at our kitchen table, we had a sense that the time was right for us. We were both soon to turn sixty, both in good health, feeling financially secure and always up for a challenge. The prospect was too much to resist so the decision was made from our hearts to "have a go." This is where the head started to think harder – how to let go of the farming activities, the agricultural and agritourism businesses we owned and managed, what to do with our house, the family pets, and what of our connections with friends, family, and local organizations with which we had shared our lives for the previous fifty years. The hearts screamed - commit to the volunteer assignment then manage the rest.

The decision to volunteer:

The decision was made within hours to "go for it." Managing the consequences lead to a busy few months. News quickly spread to family and friends, sometimes met with tears, others with joy, but always with amazement. In the words of one incredulous friend "if anyone was ever going to do something completely ridiculous then it would be Pip and Simon." Maybe we were right for an assignment in Ethiopia!

We were to learn that AVI has programs in many countries of the Pacific Region, Asia, and Africa. Some relate to health, others education. There are engineering roles and people working in marketing, health, promotion, fundraising and a number in agriculture.

The AVI East Africa program is based in Tanzania, but they also had a country manager in Ethiopia. We gradually ascertained that our personal safety, the provision of all the basic human needs and any physical or psychological support would only be available from these people. Time would ultimately show that the recruiting choices made by foreign affairs staff in Canberra and the local embassies were excellent; we developed very close relationships with our local contacts: Tanzania-based, Musa, and Ethiopian local, Demiss.

The easy part was to find Ethiopia on a map. There it was sticking out on the horn of East Africa – at least we now knew something!! The hard part was trying to come to grips with the extraordinary history of the country and its unique cultures and languages.

Preparation and training:

With AVI pre-departure workshops behind us, and arms full of vaccinations, suddenly it was just a couple of weeks until we would fly first to Arusha in Tanzania for a week of cultural and security briefing and then on to Addis Ababa in Ethiopia. But what to pack? Think of the essentials of life we reasoned. Clothes would be available locally, but would there be bedding – or a bed? Pack the sleeping bags and sleeping mats. What of the water supply? Surely it is likely to be very high risk to drink and cook in local water, so we were given a heavy but reliable twenty litre water filtering system.

Presumably local food and cooking facilities would be available, but Australians could never head off for two years without a good supply of Vegemite. It would not be long before we discovered such delicacies could become real comfort food when a very long way from home.

But our whole future life needed to fit in two suitcases!!

We were careful not to dwell on the possible medical issues which could confront us over two years, but we updated our first aid accreditation and packed a large first aid kit. Hey doc – what is a "sharps kit"? That is what you take when you may need to be operated on somewhere that there is no sterilized medical equipment – scalpels, syringes, sutures, antiseptic. Really? We took two kits!!!

And so, with a mix of excitement and trepidation we were off to Melbourne airport laden with exactly thirty kilograms each of luggage and bulging "carry on" bags and in our hearts a feeling of deep compassion, but in our heads a wealth of knowledge. Pip had been a Farm Advisor for many years with enormous experience and qualifications in aquaculture, dairy, sheep, pastures, agroforestry, ornithology, and climate change. Having been trained as an Agricultural Economist, I had been farming for forty years and with Pip's help, had developed businesses in aquaculture, agritourism and farm forestry whilst operating our sheep, cattle, and cropping farms.

Arriving in Africa:

The sprawling sea of commerce at the Dubai airport became the first meeting point with other volunteers also beginning their two-year assignments in Africa. There we met up with two young women – both headed for assignments in Tanzania. Little did we all realize at that time how close and dependent on one another Australian volunteers become. We work in different roles and live in different villages, towns, or cities – and even in different countries - but ultimately all volunteers share an experience which creates close and caring relationships.

Landing in Arusha brought back many memories for Pip who had spent four months in Africa many years before, and she had always harboured a desire to return. This was the beginning of that opportunity, and she and I soon became fully engaged with the week of training in culture, moral dilemmas, language, safety and security, and the unique positions that volunteers find themselves in developing nations. I turned sixty on the day we flew out of Tanzania headed for Addis Ababa. Was it a day of excitement? Certainly. Of trepidation? – a little. A flight surely into the unknown.

AGOHELMA – A truly local Non-Government Organization:

The Abebech Gobena Child Support Agency (AGOHELMA) is an organization established in the early 1980's during the period of disastrous famine which attracted worldwide attention and led to the Live Aid Concerts. At that time, the much-loved founder, Abebech Gobena, found two young babies still in the arms of their parents who had died of starvation, and she took the orphaned children for medical attention. So began one of the great humanitarian stories - of a lady who gave up all her worldly possessions for the care of ever-increasing numbers of orphaned children.

Abebech was a visionary social thinker and long before it became established practice, she worked out that orphaned children were much better supported in a community than in institutionalized care. She also determined from day one that her organization would not depend entirely on donations but would develop its own income-generating activities.

This was the organization to which we would be attached for the duration of our stay in Ethiopia. The head office is in Addis Ababa. Some of the two hundred and fifty staff tend to the needs of the small number of orphan residents (mostly those with HIV or disabilities) but many staff are involved with the income generation activities and distribution of aid to the many families who are providing homes and family life to thousands of orphaned children placed in the community.

AGOHELMA also developed a large child and maternity hospital, two large primary schools, hospitality training for women and IT training for disadvantaged young people. From the Addis Ababa compound they also provide hundreds of free meals daily and toilet and showering facilities to the seriously disadvantaged people in their area of Addis Ababa, close to the famous Merkato market – reputedly the largest open market in Africa.

In the early 1990's, the Ethiopian Government gave the Abebech Gobena organization a forty-hectare farm in the West Shoa region, at Birbirsa, close to the village of Guder, one hundred and fifty kilometres west of Addis Ababa. At this location, in the beautiful lush green highlands of Ethiopia, AGOHELMA established a health clinic whilst developing the farm. This was subsequently used as a base for the establishment of a number of primary schools and more health clinics in the west Shoa region which until that time, had seen little government support or interest - other than in times of tribal conflict. As these facilities became successfully established, they were handed back to the Ethiopia Government.

The farm was to become an income generation activity and not just a base for other community support.

Our role and challenges

For some years, the costs of running the farm had far outweighed the income generated. We became aware that the farm was overloaded with managers who previously had roles administering the provision of health and education services in the area. There was a purchasing office, a cashier, an accountant, general managers, deputy general managers, overseers, and managers of dairy and crops. Almost more managers than could stand upright on a forty-hectare farm!!! Could this be reversed?

This is where Demiss Feyissa enters the picture. He had spent years working in the aid (NGO) sector in Ethiopia and had a long-standing relationship with AGOHELMA. As manager of the small but growing AVI program in Ethiopia, he was looking to develop assignments for Australian volunteers. Between he and the AGOHELMA general managers they proposed an AVI assignment to try and turn around the financial situation of the farm.

Initially they considered that the role would be based in Addis Ababa with regular visits by the volunteer to the farm and developing reports for the farm managers. Demiss was apprehensive that volunteers could not be found who would be prepared to live in the very basic accommodation available at the Birbirsa farm. The Australian Department of Foreign Affairs and Trade (DFAT) officials in Canberra were very concerned at the prospect of having volunteers in their program living remotely, and in one of the most politically insecure areas of Ethiopia.

It was clear to us that to effectively do the role as farm management mentors we would need to live on the farm and work alongside the farm staff.

For the first three weeks of our stay in Ethiopia we were grounded in noisy, sprawling Addis Ababa receiving more language and culture training and security and health briefings. Meanwhile Demiss was busy working with his AVI manager, Musa, based in Arusha, Tanzania trying to get clearance from DFAT in Canberra and from the local Australian Embassy in Ethiopia for us to head to our new "home" at Birbirsa.

Life in Ethiopia – a daily record

To tell the story of the next two years I have taken extracts from our daily notes. Every day brought moments of excitement and drama, but there were also moments of a simple daily life, and others of frustration, emotion, hardship, joy, pride, and humility. If there were moments of boredom, they were rare as there never seemed to be a "dull moment" in this extraordinary country. These notes were a raw reflection of daily life, usually written after we had finished our simple dinner in our very simply hut. Sometimes the power would be on, but very often we would type by the light of a candle, relying on the battery of our trusty laptop.

Often, we had medical discomforts associated with living in such basic conditions and our water supply was intermittent. Our kitchen was on our small veranda, and squat toilet a few hundred meters away. But with no phones, no television, and no internet available near the hut, my evening entertainment became writing and Pip took to needlework. Our personal relationship and marriage had always been strong, but it became stronger by the day – we were almost in each other's pockets for two full years and totally dependent on each other for daily support. So often one would be down – coping with frustrations of homesickness, unpalatable food, no water, intermittent (or non-existent) electricity, physical injury or acute gastric upsets – but inevitably the other was up for the challenge and provided support. Our emotional tides washed in and out, but always our hearts were strong, and our focus never wavered from the desire to help people so desperately in need, and on the work to be done.

I am sure that throughout this book there will be people mentioned who may have a very different recollection of events that I write about. Others will think that I have been overly critical – or not critical enough - of people, places, and cultural practices. To those people I apologise. The words here have not been written after a period of reflection but were written "in the moment" – sometimes of frustration, elation or whatever the emotion was at the time. Some of the "facts" that I write about may also not be accurate as I have written at the end of each tiring day, attempting to reflect what I had been told and observed – always from memory and always told to me by people with English as a second or often third language.

The boxed text is written in retrospect, but the rest was written in the moment.

In the final pages of the book there is a brief introduction to the key personnel who were instrumental in our assignment. All of them were of great support to us and have our sincere thanks for that support and friendship.

Chapter 2

Jan 2014 - Settling in Ethiopian Style

EE.1: AGOHELMA farm at Birbirsa, Ethiopia:

Our first opportunity to see the farm was a day long excursion from Addis to the farm at Birbirsa. This gave us the chance to purchase the basic furniture and supplies back in Addis that we thought we would need for our hut when we shifted permanently to the farm a few days later.

Day 25 of our assignment
Our first day at the AGOHELMA farm

A big day full of lots of experiences. An early start and on the road for our first visit to the farm with Musa, the AVI East Africa manager (based in Tanzania) and Demiss our local AVI manager. Travelling through Addis Ababa, the roads seemed quite modern with some flyovers and roundabouts, and all the way for the 120 kilometres to the town of Ambo and so on to Guder, the road surface was amazingly good.

The sides of the road through the suburbs of Addis and the cities, towns and villages along the way are of course vastly different, with people and animals walking everywhere. We passed thousands of little shops and businesses, and even a stock market outside a large abattoir, where small mobs of ten to twenty sheep and goats stood in tight formation with their heads facing in – disciplined with a light cane by their handlers. On the edges of Addis there are large plantings of Australian eucalyptus trees introduced into Ethiopia in the 1800's to provide firewood, doing their best to suck up the diesel fumes and dust that are so much part of Addis Ababa suburbia. At this time of year being the dry season, it is clearly a losing battle.

On the outskirts of Addis, we picked up the general manager of the AGOHELMA (Ato Eshetu), a large affable fellow with a big, round face and keen to see the farm improve its profitability with high hopes for what we may be able to achieve.

From left – Pip and I, Musa (AVI manager East Africa), Demiss (AVI manager Ethiopia), Ato (Mr) Eshetu and Ato Tesfaye (senior managers of AGOHELMA organization) – at the head office in Addis Ababa.

Once away from the seething mass of humanity in Addis Ababa and heading west, the landscape becomes endless rolling hills and valleys dotted with thousands of small farming paddocks creating a massive patchwork quilt spread across the countryside. At this time of year there are tiny paddocks of harvested teff and barley crops and areas of blue gums, mostly in dispersed clumps or blocks. Some juniper trees and native acacia trees are scattered about, mostly close to the visually enchanting 'tukuls'. These are the local dwellings: round huts made of timber and mud with a roof of brush. Sadly, from a visual perspective they are rapidly being replaced with square mud and cement huts with corrugated iron roofs. As we chatted expectantly it really began to sink in that we were living the dream.

What would we find at the end of the rainbow when we arrived at the farm in the village community area of Birbirsa?

We had two and a half hours driving to get to Ambo, a town of about sixty thousand people, many of them students at the local university. We stopped for lunch at the Abebech Hotel, a six-storey concrete block towering over the tiny shops and markets that make up the Ambo township. It is set in lush tropical gardens - such a contrast to what we had expected of Ethiopia just a couple of weeks ago. Surely, we had reasoned, we were heading to hot and dry dusty plains with flat topped acacia trees. Nothing could be further from the reality.

Demiss introduced us to Mohammed, the hotel manager, as we are told this hotel will be our main link to the western world, with comfortable beds, hot showers (sometimes) internet (occasionally) and a toilet to sit on - rather than squatted over. The hotel is to be our "Safe house" in the case of an emergency in our farm area, which is still another twenty kilometres further west.

We have been told that DFAT want us to take a "rest and recreation" weekend away from the farm every month and, as it seems a very long way to Addis, we may end up being at this hotel quite regularly. Apparently DFAT are a bit worried about our posting and we have been given walkie talkies so we can keep in contact with each other whilst we are at the farm.

After lunch we headed to the hospital in Ambo to check out our medical options if required over the next two years. This was our biggest culture shock to date – and we felt that Demiss and Musa were also shocked by what we found.

We waited in the dirt courtyard under some large rotting jacaranda trees, watching people fill water bottles and wash their hands and faces from the tap in the middle of the courtyard. We suspected that most were the family of the sick whom they were looking after in the "wards". A few visibly sick and injured people were also walking about.

We met the hospital director - a young Ethiopian doctor who was from Ambo before he did his medical training. He was not very inspiring - but what resources did he have to work with? He took us along outside covered walkways with billboards and notices relating to AIDS and HIV to the surgical wards. The wards were horrific.

The cement on the walls had been painted decades ago and was peeling off in places. But nothing could compare to the sight of the mattresses, most with torn covers and the sponge interior harbouring "who knows what" fluids and diseases. Sheets, pillows, blankets, equipment – all non-existent. Hard to imagine how anyone can be treated here. For surgical operation - a bare table. I guess, at least it is easy to clean.

In the corner was a fancy machine all covered in plastic but by the look of the dust that had settled on it, I doubt it featured prominently in surgical procedures carried out on those people unfortunate enough to be brought through the front gates of this facility. A rack of small white bandages was hanging outside the ward door, and a large orange plastic bag for all infectious/waste material.

Otherwise - nothing. The green surgical aprons, which I presume had been washed with a hose after the last operation were lying in the sun on some grass and others hanging on a fence out the back. It flashed through my mind that the front-line surgeries in the first World War which we see depicted in the movies would be preferable to being treated here.

We resolved not to ever go to this place again, but to go straight to a private hospital in Addis even if it is three extra hours in the vehicle. Virtually all pharmaceuticals can be bought at drug stores with no need for prescriptions, so hopefully we can get some decent antibiotics if we need them. We understand that there is a new hospital being built on the Ambo university campus which probably accounts for why this facility had fallen into such a state of disrepair.

We then proceeded another 12 kilometres further west on the main road heads to the Sudan border in the Gambella region. In the busy little market township of Guder, we turned left off the main road onto a rough and rocky road which climbed gently for the seven kilometres to the farm at Birbirsa. Never a moment without soulful donkeys and their handlers on the road loaded with firewood, bags of grain, charcoal, clay pots and other local produce heading down the road to town - or returning up the road with basic household requirements or containers of water.

We pulled into the farm through large steel gates opened for us by a guard (one of eight employed by the farm – four day and four-night guards!). We were met by the farm manager, Tiberbu, a kindly man in his fifties in a safari suit who looked a bit like a schoolteacher. The deputy manager, Umdetta, was also present, about the same age and fairly quiet.

Our entourage were ushered into the office which was one of eight rooms facing a grassy courtyard – obviously the centre of operation for the surrounding farm. An old computer and printer were on the desk, with a map of the farm, a filing cabinet and two benches to round out the office furnishings. Tibebu gave a brief talk, then we were shown a choice of rooms for our accommodation quarters.

The first was in one of the older mud brick buildings around the courtyard. The room had two single stretcher beds, a low reed matting ceiling and a door into another dark little empty room – at least that is what we think it has as there are no windows and gloom and darkness crept out of every crevice!!

The other choice for our home for the next couple of years is two rooms in a slightly newer cement building, currently the farm manager's offices. The rooms are not connected by a door but both open out onto a small veranda and then the courtyard, and they have steel doors and glass in the windows - quite a rarity we discovered as all the local huts have wooden shutters instead of windows. Each room has a single light globe and in one, a power board dangling from a couple of electric wires. Our "kitchen" will have to be on the veranda we have decided. We have chosen those rooms as our future accommodation. Definitely unfurnished and very definitely basic, but we are sure that we can soon make them into our home.

Beginning to unpack our worldly belongings at our hut. Living room left door, bedroom to the right and kitchen on the veranda.

The water supply is from a single tap the other side of the courtyard, about twenty meters away from our hut and the toilet shed 100 meters away behind the courtyard buildings. The toilet shed has four separate cubicles, reasonably clean and each with a hole in the cement floor for squatting into the long drop.

From there we went for a tour of the farm – first, a little mud building housing the dairy processing area with a lady squeezing water from freshly made butter with her hands. There was the separator, and a butter churn and barrels of fermenting milk for cheese. Then on to the oil processing shed where two men were feeding bucketful's of niger seeds (an oilseed native to Ethiopia) into an extracting machine, with beautiful fresh oil flowing out one end and dry oilseed cake coming out the other.

The farm also has a stone wheel grain grinder which they offer for use by the community. We were shown a new cattle barn which was being constructed, and then the old barn with about fifteen dairy cows tied to the troughs inside. Behind was a wire shed that had been a chook shed but is now empty. Then on past some large yards with covered areas which had been used as a feedlot but now house the eight farm working oxen. We then inspected our corrugated-iron shower cubicle. All cold water of course and a couple of hundred meters from our hut. Such an enticing prospect for a daily wash!!!

The main issue that organization manager, Ato Eshetu, wants us to resolve is that of the irrigation pump. Some years ago, a bore was drilled over 100 meters deep with some Aid money, and a submersible pump had been installed. It has been sitting there for more than five years and has never operated. The electrician says it's the pump and the pump supplier says it's the power supply. Tibebu rightly points out that being able to irrigate part of the farm would greatly improve the profitability. It sure sounds like a fun project which could make such a difference, if we can get it going.

Back to the courtyard for a coffee which a young woman had been brewing ever since we arrived. Strong and sweet in tiny cups - coffee had never tasted so good!

We were tired but looked forward to the possibilities ahead and had plenty of discussion as we piled back into Demiss's van and returned to Addis.

EE.2: Arriving at the farm to begin our new life

A few more days spent in Addis Ababa with numerous shopping trips and more training – but an ever-increasing urge to get to the farm and get our teeth into some action.

Day 7

After a long morning outside the Kaleb Hotel in Addis Ababa packing all our gear into Demiss's van and the farm Hilux ute we were finally off headed for Ambo. On the outskirts of Addis, we stopped at one of the may fruit and vegetable stalls to stock up on bananas, avocados, potatoes, pumpkins, and eggs.

Driving on, and seeing the landscape a second time: the patchwork of tiny fields, some ploughed, others lush and green with crop and always a few mobs of sheep, cattle, or goats scattered over the rolling hills watched by young shepherds. There are donkeys loaded with bags of grain, often trotting in a group, with a man running way behind. Pip is already fascinated how they tie the loads onto the donkeys.

We watch masses of people as we pass through the villages and on the road between the villages, always walking with plastic bags full of local produce or holding sticks across their shoulders, and herds of goats, soft brown zebu cattle, and the occasional farmer with his wooden plough over his shoulder driving his oxen to or from his home after tilling yet another field. Depending on the time of day, masses of school children in uniforms are walking along the road to or from school. The large, flat topped thorny acacia which defines the African landscape, gives great character to the green rolling hills.

We arrived with our two vehicles packed with gear at the Abebech Hotel in Ambo around 6.00pm. There we checked into a hotel room on the fifth floor - large, spartan and worn out but a good view of the town. We went for an evening stroll down the main road with Merka (the farm ute driver) and Demiss, along with another few thousand locals – all out for an evening stroll. We were shown the two ATMs in town which were to become our source of finance as we could draw cash (Ethiopian birr) from our Australian bank accounts.

EE.3: To market, To market

Language was always going to be a problem and sadly we never became fluent in either Amharic or Afara Oromo.

The national language is Amharic which is mostly spoken in Addis Ababa but there are another eighty distinct languages spoken in the country.

The language of Afara Oromo is commonly spoken in the countryside, as about thirty five percent of Ethiopians identify as Oromo and Oromia is the largest region in the country. Most of the farm workers spoke only Afar Oromo, however the managers all also spoke Amharic and some of them also a little English.

On the occasions when the managers from the organization head office came to visit the farm from Addis, they could only speak Amharic, so they communicated with the farm managers but not the farm labourers. As we learned, one of the many unique aspects of a very unique country.

Day 45

Monday is livestock market day for Guder. It is the biggest stock market in the west half of Ethiopia. We had talked about buying a sheep for me to kill so we could eat some decent meat, and the management were keen to take us to the market.

We bundled into the ute (six of us as usual including Tibebu and his deputy, Umdetta) and off to Guder at about 9.00am. The road down gets more and more crowded as we get closer to town with donkeys and women carrying huge loads and men and boys driving anywhere between one and twenty sheep or goats in little mobs, plus numerous donkeys, and some horses. Other boys walking briskly in the Guder direction, carrying one or more live chickens.

The farmers come from every direction across the farmlands and mountains, beginning their journeys days before the Monday sale and slowly they all converge on the chaotic roads that lead into the Guder marketplace. We wanted to get a few bananas and oranges and thought that would be a good activity before buying a sheep, but the management predictably decided that it should be the other way around, so we parked the ute and set off to walk to the market.

I am starting to get the feeling that the word amazing is going to get quite monotonous. But here it comes as it is the only way to describe the market. AMAZING. A hillside completely covered with a seething mass of humanity and animals. Goats, sheep, bullocks, and donkeys. The larger animals were controlled with ropes around necks or had their front legs hobbled, and the sheep and goats just shepherded in small groups by men or young boys. We could not possibly walk about without pushing animals and/or people aside. It was completely jammed, an area of five hectares at least. It wasn't smelly or overly noisy or dusty, just extremely crowded.

We are in there somewhere purchasing a sheep.

Fortunately, by now there were just three of us - Pip, me, and Umdetta, as it was hard enough for us to stay together. I found a nice young ewe and the bargaining began but Undetta could not get the seller down below 1000 birr (A$50) which Undetta considered excessive. Eventually after pushing and shoving our way through the crowds I found another nice young female sheep for 750 birr (A$ 38). I had brought a collar and chain, so we put that on the sheep and with the help of two young boys started dragging our purchase to the gate through crushing crowds.

When we got to the gate a man said we had to have a receipt to get out of the market area, so Undetta headed back through the crowd to try and find the seller. As soon as he was out of earshot, the man decided we did not need a receipt. Sometime later Umdetta reappeared clutching a receipt (which I will treasure forever as a memento of our first sheep purchase in one of the most extraordinary places I have ever been in my life).

Proudly we led our very well-behaved sheep a kilometre through the streets of Guder back to the ute and chained her up in the back. Then of course we had to do the rest of the shopping. We had seen scores of men standing in one area of the town selling individual chooks and roosters and thought it would be good to have one to eat so, again with Umdetta's help, we negotiated a price of 120 birr (A$6) and bought a splendid young rooster with its legs tied to prevent escape.

As we headed back to the ute, we could not find bananas or oranges in the main vegie market (today mostly just tomatoes and chillis for sale). We linked up with the other managers and headed for home up the bumpy track, with the sheep tied standing in the back and the rooster in the front!!! We spotted a lady selling local bananas from a tiny stall beside the road and although they looked almost black and inedible, we bought all she had and actually, they were delicious.

When we got home, we unloaded our sheep with many helpers who now all decided she was pregnant and so I could not possibly kill her for us to eat. Of course, Pip was easily persuaded, so now we have a lovely looking ewe tethered in our courtyard keeping our "lawn" down. Heaven knows what I will do when or if she has her babies!!!! Already she seems to be the farm mascot, and everyone keeps untangling her rope. Formally now known as "Mrs Sheep" to all and sundry.

Much to the amazement of the local people, I killed and plucked the rooster, a skill I had learnt from my grandmother and father decades ago. It is now in our fridge and will be roasted for dinner one night.

We had returned from Guder at about 1.00pm and the managers had a good rest and then it was decided that a coffee ceremony should be held so we sat in the courtyard yarning from about 2.30 until 5.00pm whilst the cleaner squatted over her little charcoal stove and produced lovely little cups of coffee.

Our Amharic is improving slowly but their English is improving rapidly, so I noticed today that we all spoke English nearly all the time. We had the manager, deputy manager, overseer, one guard, the accountant, cleaner and driver all tied up for the whole afternoon, so I hope something was achieved!!!

Afterwards Pip and I went for a short walk and to see the new calf that had been born last night. We have made some good progress with management changes, and everyone is delighted that the milk production has gone up from average 6 to 8 litres per cow each day thanks to the better feeding and water changes that we have initiated. Pip is frustrated because there is so much more to do which we know will make even more progress, but I am sure the management can only handle so much change at a time and so we need to try not to push too hard - even where animal welfare issues are involved.

The next problem with our sheep purchase, is that she needs to be inside at night or will be killed by wild dogs or hyenas or stolen- even if kept in our courtyard. Management has decided there is a spare room in the courtyard building that she could occupy, so a key was found for the lock on the door, the cobwebs and about 2 barrow loads of termite dirt taken out, and she is now asleep in a padlocked room. In Oromo language – the Hola Manor (sheep house)

We are managing to find some good physical exercise each day whether it be digging the veggie garden or removing woody weeds, walking, or playing with our Frisbee and footy. We have no problem with the mental exercise as we are adjusting to new challenges every day - working out how to implement our management changes, learning the languages (Amharic spoken by the managers and Afar Oromo by all the locals) and researching the information we need to brush up on in our "new" agricultural environment.

EE.4: Social Issues – polygamy, HIV/Aids, alcohol

Since the farm managers often had very little constructive work to do, we had the wonderful opportunity to spend time with those who spoke some English, loving our conversations about issues affecting the local community and Ethiopia as a whole. These discussions often took place over countless cups of traditionally brewed coffee or around the campfire we frequently lit in the evenings outside our hut.

Day 48

Up at 6.30am to go to the dairy shed, as we have decided we will record the milk from each cow every Thursday morning and afternoon, to see what effect our changes are having. Hopefully we may even inspire management to ensure that it happens in the long term. Tibebu appeared about 8.30am when we were nearly finished so another good chance to share our ideas with him at the coalface and not in his office.

We get on extremely well with him with lots of jokes and his English is becoming more fluent, so we can ask him about anything and start a good discussion. We call the rich men of the area the "Kings" and Tibebu tells us that one of them (whom we have met) has 12 wives and more than 27 children. This has led to an interesting discussion about polygamy and where it fits in the culture, and lots of discussion about religion.

Today he told us the history of the farm and lots of information about HIV and AIDS in the area and all the work that AGOHELMA has done over the years on these issues, providing support and treatment facilities here at Birbirsa, in Guder and in Ambo. He says that prevalence is well down in the last few years but there is always a fear of it increasing again. He says there is a direct link between alcohol abuse and sexual violence which leads to more AIDS. Alcohol abuse is increasing anywhere within a few hours walking distance of the larger towns, hence his fear of AIDS and other sexually transmitted diseases increasing.

Back to the hut for a late breakfast and then to check emails, sitting on our special seat out in the paddock where we get internet connection (sometimes). We found everything had got to the right people for our submission for an Australian Government grant to build a chicken caravan and develop an egg production enterprise. We also wrote an email to the Australian Embassy in Addis to tell them how safe, secure, and happy we are with our assignment.

Pip made a cake (packaged) in our "oven" which turned out perfectly. When the power is on, the oven consists of a really big saucepan on the electric stove. Inside the saucepan is an upturned ham tin with a small sheet of corrugated iron on top. We put the lid on the saucepan and a towel on top, and it all gets hot enough to work as an oven for baking bread, cakes and hopefully roasting a chicken. Our stove on the veranda has an electric hotplate and a gas ring so we can use either. When the power is on, we use the electric ring which is slow but saves our gas.

Fairly long afternoon nap for me as I have another head cold, getting progressively worse. I hope it is on the improve tomorrow. Pip went for a late afternoon birdwatching walk. Being alone, she took the 2-way radio with her that we were issued by DFAT for use when we are separated on the farm. She called me a couple of times, and it seemed to work OK.

EE.5: Going to town

Day 55

Up early again and into the cow shed by 6.30 am to record how much milk each of the cows are producing. Hopefully when we prove the value of this exercise, it will become the responsibility of the dairy manager, Getun. Two ladies milk the cows by hand and start at about this time each day. The cows have been tied up at their stalls in the shed for the night, but we are dismayed that they get no water from the time they were brought in by the shepherds last night - around 5pm - until they are let out after milking in the morning and taken out for a drink and grazing the pastures. We are already thinking that we can get some huge production gains by working out how to ensure the cows have water to drink overnight.

The afternoon milking begins at about 5pm, after the shepherds bring the cows back in from the fields and tie them up in the sheds.

After the milking was finished, we headed back for breakfast and were told that there was to be a trip to Ambo. To get to Ambo we travel down a very rough gravel road for seven kilometres to the village of Guder and then another twelve kilometres along the main bitumen road which heads out a further 600 kilometres to the west of Ethiopia where it borders Sudan. Ambo was a village just a decade or two ago but is now a town of over sixty thousand people, including around nine thousand students at the university.

We got our morning jobs out of the way and sure enough by 9.30am we were on the road with "only" five of us in the ute. Apparently one of the buyers of the cows had paid his 18,000 birr (A$900) and the money had to be taken to the bank. This has to be done by Kidist, the cashier. Unfortunately, Tiberbu has been selling some of the best cows to raise funds to keep the wages paid, which will make our job of turning the financial situation around more difficult, but we will just have to work with that.

Here is a description of a trip to Ambo:

Today we went to Ambo in the farm ute. Really, we only needed to buy a cabbage but got a few other bits and pieces. By the time we got home a few hours later, I said to Pip that the feeling I had reminded me of when I was about ten years old, on the way home in the car from a family trip to the Royal Melbourne Show: total exhaustion with the mind still spinning from everything it had processed – but this time, fifty years later, it was just to buy a cabbage!!!!

If we could have a dream it may go like this – "we would be to be left in peace to walk down through the market, assess things and make the odd purchase, then meet up with the managers in an hour or two".

The reality is – the farm driver Merka drives us around the town, and we drop managers off here and there. We presume that they will need to be picked up again in a few minutes but may not appear for a few hours (so we never know whether we are in a hurry or killing time!).

Then we get "driven" though the market in the ute when it would be so much quicker to walk, as it is completely clogged with traders and their produce. Along the way we pass lots of shops that may have items on which I would like to do some research (poly pipe, hose, nails etc.) but no - we are heading for the cabbage area. Mostly the ute is stationery or crawling, tooting at people and donkeys. When we get to the vegetable area, there is usually nowhere to park as the whole area has developed over recent centuries just for people and animals and not parking spaces for Hilux Utes. We could easily have parked on the main road, only a couple of hundred meters away.

When we start the purchasing process led by one of our managers (usually with the advice of two or three others), we are not sure if they are negotiating on our behalf or if they are buying for themselves. Sometimes, when we think the deal is done for a kilo of tomatoes for say four birr (20 cents) we are dragged away because the price is too high and so start all over again at another stall hoping to only pay three birr. (By this time, I am usually ready to pay ten dollars just to get out and head for home).

Eventually a few things are in our backpack and as always, we are completely out of change, as some of the stall holders cannot even find the change for four birrs out of a ten birr note (50 cents) - but the

ATM we use in Ambo only spits out one hundred birr (A$5) notes – when it is working.

Back into the ute that Merka has been guarding whilst we are away with the managers. We dump all our purchases on the seats and floor where it will share leg and bum room (never good for the bananas or avocados) with a huge crowd of bodies as we return to the farm, because anything may be stolen from the back of the ute.

Slowly we extract the ute from the crowds in the market and drive a few hundred metres back to the main street. Then into a coffee house where we wait for an indeterminate amount of time for others to appear. They coordinate their movements by mobile phone - if the phone system is working.

We then all pile into the ute and head for Guder. On the outskirts of Guder there is another stop at the coffee stall in a little tent next to the hot and dirty tyre repair shop. Although the coffee is always delicious, by now it is the last thing I feel like drinking as we have had coffee a few minutes before in Ambo and just getting everyone into and out of the ute is tiring enough.

Having been introduced to whoever is in the coffee stall and had lots of laughing about who will sit where, we consume a couple of cups of coffee and then it's back into the ute. Usually, I take a look at the church spire over the road and I pray. Always the same prayer - that we are heading straight home - but mostly there are another one or two stops where we wait in the ute with feet on bags of fruit and veg and knees wedged under the chin.

As we start to head up the road we are asked if there is anything else we need. By now it would need to be seventy virgins (like a good jihadist martyr) for me to say "yes, let's stop again"!!! The last seven kilometres up our road is so rough, all resistance to discomfort has been thoroughly extinguished.

We were back for a late lunch then a good sound sleep to recover from the morning. Later we headed back down to the dairy to measure the milk from the afternoon milking. Then a bit of pick and shovel work for me in the area where I am determined to create some evaporation ponds for the effluent from the biogas plant.

This plant was installed as part of a project sponsored by the Ambo University many years ago, but it has never operated. Another good project for a couple of Aussie volunteers! Still no gas emitting, but I will give it another week then have a test to see what is brewing in the large dome. I did quite a lot of research about biogas plants on the internet and found that it is important to get the correct mixture of cow dung and water put in each day, so we now have the cow staff getting that right.

Chapter 3

April 2014 - Customs and Traditions

EE.6: Easter - Life and Death

One of the great joys of living right in amongst the community was the opportunity to share in the everyday life of the community and we relished it. So often we had no idea what was to eventuate throughout the day and with our limited language skills we just went with the flow..........

Day 65

Last evening Pip and I walked the 7 kilometres down the road to the local town of Guder. We spent the night in a school compound with an American couple who are living in Guder on an American Peace Corps assignment. Terry and Ellen were both teachers in America and came to Ethiopia a little before us to teach at a local school. Terry made yummy pizzas for our dinner and being Easter, the night was pretty quiet considering the locals were all partying on meat, araki - the locally made spirit - and tella - the local beer, having just broken their pre-Easter fasting.

Breakfast with Terry and Ellen consisted of fried potato chips, scrambled eggs and coffee, then lots more talking, sharing Ethiopian experiences and politics from our own countries until lunch time. We all walked about two kilometres into the centre of Guder, through a new 'housing estate'- a sloping paddock with a number of timber and mud huts being built with new (very thin) corrugated iron roofs and about fifty metres between each hut. Ellen walks this way to work every day so she must be pretty fit by the time she carries their supplies home from the market after she finishes work.

We had our special "Easter Sunday" lunch at the restaurant in town – you guessed it – injera with lamb tibbs as the menu is only in Amharic, the waiters speak only just enough English to make you think they may understand but invariably when the meals arrive, they clearly had no idea. The injera with tibbs is the only thing any of us can enunciate sufficiently in the local language to get the message across.

On this occasion we ordered injera with three tibbs (bowls of tough lamb meat in moderately spicy sauce) but of course only two arrived but that was enough. After lunch we all walked up to the other end of town and found a "café" open that had stale donuts and coffee available, so we availed ourselves of those luxuries to finish our party lunch.

Pip and I then set off for the uphill walk to the farm. It was pretty hot, but we had water with us and we powered along well. We were joined almost all the way by varying numbers of children – some just enjoying walking along with us and others asking for money which gets trying but we are getting better at handling it. Due to the excesses of the previous night and morning there were almost no adults to be seen on this trip and absolutely no traffic. We had been warned that the buses would not be running.

You would think that our Ethiopian episodes may well be over for Easter at about 5.00 pm Sunday night. But not to be. As we were having our cup of tea, the farm overseer, Lechisa and his brother who is a nurse and is visiting the area for Easter, came into the courtyard and headed into the health clinic in one of the rooms in our courtyard. It is run by the local "community nurse" and she comes to the clinic occasionally to administer vaccinations to children and contraception to the women. I did not previously know, but obviously Lechisa has a key.

I noted that Lechisa had a white gabi (a large white cotton blanket) over his head, and he said he was feeling unwell. I made a few jokes about him having too much raw meat and tella (local beer) and his brother asked if I had any test strips for identifying malaria - which we did not.

As I was about to head off, he told me that his sixteen-year-old half-brother had been accidentally killed with a spear on Friday afternoon. Apparently, he had gone hunting with his younger brothers and somehow the spear hit him in the heart and that was it. The family, including Lechisa, had walked for hours to get to where it had happened. Now it all gets a bit complicated. Lechisa is the son of Angussa, one of the guards here at the farm.

It turns out that Angussa has two wives (which our farm manager Tibebu did not know), and the son that was killed was from a wife (number two) who lives in the nearby village of Dogoma. Angussa and wife number one live just down the road from the farm here at Birbirsa and we socialize with them quite a lot. They have lots of beautiful children of which Lechisa is the eldest.

Lechisa in his family compound with some of his brothers and sisters.

Lechisa told me that the funeral would be on Monday, and we should come to Angussa's house at about 2 -3pm on that day. He and his brother headed off with a handful of pills from the clinic. We are told the health clinic has no role in treatment of disease or injury and is just for preventative activities.

I made fish cakes for dinner from grated potato and a can of tuna. Pudding of coffee and cardboard flavoured custard from a packet, and banana. Being a special day, we had one carefully rationed piece of Toblerone chocolate as our easter eggs.

EE.7: Funerals in Ethiopia: A profound social tradition

With a large population and relatively short life spans, it is not hard to understand why funerals are a big part of life in Ethiopia. Pip and I rarely heard about the birth of a child although of course it is a very common occurrence with the average number of children per family being more than six in many rural areas. Funerals, however, are by far the largest social gathering, and it was not long before we had become quite experienced with the etiquette of the occasion.

In most cases the day after a death in the community there will be a funeral procession from the home of the deceased to the local burial ground which is often in the yard of a local church. For the very poor, they will be buried in a plot behind their home.

Depending on the importance of the deceased – or their family to the community, the procession will range from hundreds to many thousands. Most of the close relatives and friends will be neatly dressed in a white gabi over their shoulders for the men and over the head like a scarf for the women. The body of the deceased will be lying on a stretcher or old couch and completely covered with their gabi.

The stretcher will then be carried by a group of men who will alternate in duties as pall bearers, as it can be many kilometres to the cemetery. The burial takes place and the assembled masses stand around for a good chat and then wander off over the next few hours.

Then begins a week-long period of mourning in the hut of the deceased, or an adjoining tent which is erected for the purpose, and many local people call in for a short visit most days of the following week to show support. Generally, local women who are close to the family provide food for the visitors from first thing in the morning to last thing at night. Those of us with some financial means hand over a small financial contribution to the family of the deceased.

The closest of the relatives then wear black mourning clothes for months – or even a year. We are told they only stop wearing black when other relatives tell them it is time to move on.

The importance of attending funerals also ties in with the Ethiopian justice system as one of the punishments which may be handed out by the head of the local kebele (village group) is to be banned from attending funerals for a certain period of time.

Day 66

Today was very quiet around the farm as most the staff had the day off and the ones that were here were very subdued and everyone is obviously preparing to the funeral. We could hear mass wailing (ululating) coming from down at Angussa's house as crowds started for gather.

At about 1.00pm we heard more chanting and wailing and looked out through the farm steel gates onto the road to see the funeral procession heading up our road for the seven kilometre walk to the cemetery at Mutulu. As it passed by, we wondered if we would join in, we but stood inside the farm gate and just watched. I did catch the eye of Lechisa as he went past, and I think he would have liked us to walk with him as we have become very good mates.

We had previously given it some thought but decided that walking that distance up the road in the hot sun and then again to get home may be a little too much for us. Also, we were not sure how long we would have had to spend in the sun at the cemetery and not understanding a word of proceedings. It seemed pretty heartless not to be with our dear friends at such a difficult time. We are really starting to feel part of a very extended family.

The funeral procession was a few hundred people on foot – the women, all wrapped in white gabis (special ones with embroidered edges for funerals) were mostly at the front and mostly the men just in ordinary clothes following along. There was various wailing and chanting and a group of men on highly decorated horses at the back with the riders making a rhythmic chant.

We decided not to go down to the house today, but I imagine there was a good crowd for the wake and to start the week-long mourning procedures. Presumably there is a fair bit of etiquette to sort out what part the extra wife and stepchildren play. Angussa lives just down the road in this community with Lechisa's mum – with whom he has eight children. No one seems to know how many children Angussa has with his other wife – one of whom was the lad who was killed.

Heading home after a funeral. Pip and Moti on the road close to the farm at Birbirsa.

We did a few jobs around the farm and Pip had some time in her veggie garden. I started building us a new timber chair ready for visitors. We lit our campfire late afternoon and put in some lovely potatoes in foil. While waiting for the potatoes to cook Pip roasted some of the green coffee beans, ready for grinding. Like all locals, we purchase the green beans at the market.

We have purchased a small electric grinder, but the local women use a length of heavy reinforcing rod and a wooden bowl (think mortar and pastel) to grind up their coffee beans. Later I cooked some yummy lamb chops, and the potatoes were perfect.

As we were having pudding (usually custard using purchased custard powder and our local milk) Tibebu came into the courtyard and sat with us at the campfire. He did accept a bowl of custard, banana, and fresh cream. He described it to us as "soft" food and we heard him telling others about it.

Given that every meal is eaten with fingers here, we presume that the "soft" reference was because he ate it with a spoon. That would be very rare in this part of Ethiopia. Tibebu spends some nights at the farm sleeping on an old bed in one of the other rooms in our courtyard. Most of the time he comes and goes to Ambo in the bus and lives with his wife and one son on the campus of the Ambo University. His wife is the catering manager – quite a big job I imagine as there are 9000 students living on the main campus.

We sat yarning around the tiny fire until nearly 11.00pm - a very late night for us. We have such interesting discussions about almost every topic with Tibebu, and it is wonderful for our general knowledge of Ethiopian history and culture. One of the stories he told us tonight was how, after the DERG communist government collapsed in 1990, there were a lot of ex-army soldiers around without jobs and some of them turned to thieving to keep themselves and their families alive.

Tibebu then had to post many guards on duty at the farm. He used to quietly walk around and check if they alert - or were asleep when on night duty. During one particular night he found one of the guards asleep who was supposed to be looking after the sheep, so he took a sheep away and hid it in his own property next door. The poor guard spent the next day counting the sheep over and over again and still only came to forty instead of forty-one, but he never told anyone.

He was worried about having to replace the sheep or lose his job but still did not tell Tibebu. The following night Tibebu quietly returned the sheep. When counting the sheep the next morning the guard could not believe his eyes that the number came to forty-one. Could he possibly have counted wrong the previous day? Impossible – but how.... He did not sleep on the job for weeks afterwards. Tibebu sometimes also took the shoes or guns of the guards if he found them asleep, and kept them for months, but no guard ever reported the loss!!

EE.8: The mourning process:

Day 67

Most managers and staff arrived at 8.00am and headed to the office to sign the newly instigated attendance book.

About 10.00am, Merka, the farm driver, came to find us and tell us that we should join him and Tibebu for our visit to the house where the mourning process for Lechisa's brother was taking place.

We walked down the road and then went into the main room of Angussa's hut and were given seats. There were about ten other visitors sitting in there. Angussa sat in a large old armchair chair looking sad and solemn but doing nothing – just sitting hour after hour - and we acknowledged him accordingly. Lechisa was also in the room, and he came to sit with us and chatted a bit.

Sitting on a mattress on the floor was the wife (number one) who lives with Angussa at this hut. She and Angussa were dressed entirely in black mourning clothes and Angussa even wore a black baseball cap. The mother of the deceased was not present as far as we could tell. Probably, she was sitting in her own hut in mourning, being visited by her relatives in the nearby village.

We were served injera with cottage cheese. It was a bit hard to eat so soon after our breakfast and with no drink. We were offered water, but Pip and I are reluctant to drink water that is from the local supply without boiling. I ate all of my injera and caused a ripple of amusement by finishing off Pips. All the time others are coming and going and usually as they left, they shook the hand of Angussa and slipped some money into his hand (apparently to pay for the funeral). Lechisa was suitably subdued but still keen to chat with us and explain proceedings. After about half an hour, we shook Angussa's hand and gave our fifty-birr ($2.50) donation and took our leave with Tibebu and Merka

Being the farm driver, Merka decided to wash the farm Hilux ute today. He did this by bringing the long hose up from the dairy and flooding the whole inside of the vehicle!! I know it is a bit hard to believe but it took three people the whole afternoon, and mostly they were watched and encouraged by Tibebu and also the farm cashier, Kidist. Not a bad way to tie up five people for a whole afternoon. Pip and I still have a bit of work to do to get things running properly!!! It then took three days for the ute to dry out inside!!!

EE.9: An Adventure in every sense

When Pip and I accepted a two-year assignment in Ethiopia we had little doubt that it would be an unforgettable adventure. But would the adventure be mostly on a professional basis, delving into new areas of agriculture or business that we had previously not experienced? Or would it be a more personal adventure, overcoming physical or mental hardship, insecurity, frustration of achieving so little that is measurable, dealing with homesickness or culture shock? Or perhaps it would be all of these things and more, but maybe we did not realize that it could also be an adventure in the true sense of the word when at times one has absolutely no idea how the next minute, hours, or day will play out..................

Day 71

As I sit here writing this on Sunday I have a headache, stomach ache, stools like water, legs like jelly and still feel we are the luckiest adventurers alive. What a day we had yesterday!

We thought we were going to have a quiet weekend after trying to organize a drive in the farm ute to nearby Wenchi crater, but this didn't suit Lechisa, so put it was off until next weekend.

The alternate for today was to take a walk up the road seven kilometres to Mutulu, and so we headed off with our backpack and water bottles and told the guard on the way out of the farm where we were headed.

We only walked fifty meters when a small Suzuki 4wd pulled up and asked if we wanted a lift. It is quite unusual to see a small vehicle on our road as there are usually just trucks or the odd bus. I initially said we were out for a walk, but Pip decided a lift up the hill and walk back would be good, so we hopped in. The driver spoke good English and told us he worked in the UAE embassy in Addis and was visiting his father in the highlands. In the vehicle were his sister and cousin joining in the big get-together to continue building a new hut for his father.

As we neared Mutulu, he suggested that he take us on to his father's village and he pointed to the top of a distant mountain known locally as Tulla Malke, way, way up in the highlands. We agreed to join in as it looked a likely fun adventure.

What a day full of experiences we had. The vehicle left the main gravel road about four kilometres out of Mutulu and headed up a steep dusty track through eucalypt and ensat (false banana) plantations.

After about a kilometre and a half, we four passengers got out of the vehicle as the road got too steep for the car to get up with the weight of us all inside, and we walked the last kilometre along a track which at some stage in a distant past had been made for vehicles! The driver worked his way up in the vehicle. We really felt that we were on top of the world and, given by then we were over three thousand meters in height, we pretty much were!!

Our new-found friend and host is an amazing man. Dr Fikru Deksisa (PhD) had been born and raised in this beautiful remote village of Melke - Abayi on a ridge just off the top of the mountain. He had attended elementary school locally then high school in Guder and later in the town of Ambo where he lived in a rented house with some other school mates during the week. Every Friday after school he would walk back home – fourteen kilometres from Guder to Mutulu and then another six kilometres on bush tracks up the mountain. On Sunday, his Mum would pack up food for the week, and he would carry it all back the 20 kilometres down to Guder.

When he completed high school, he studied at the Addis Ababa University and later won a scholarship to the then Soviet Union where he did his PhD in Forestry and Environment - living in Moscow. Coming back to Ethiopia, he started working for the Ministry of Agriculture and later for the Ministry of Natural Resources and Environment. He also spent time studying in Japan at Kyoto University.

Aimed at diversifying his profession, Dr. Fikru was awarded his advanced Diploma in Business Management from the Open University Business School in the UK. After several years working in different Ministries in Ethiopia, Dr. Fikru joined the Embassy of the United Arab Emirates in Addis Ababa. He speaks fluent Russian, Japanese, English, Arabic and three more local languages including his native Afan Oromo, Amharic and Tigrigna.

Now you might be starting to get the gist of why we had such an incredible day. As we went up in the car, he explained how he had instigated a few really large re-vegetation projects in the area both with eucalypts and native trees and we certainly saw plenty of plantations as we went up, and up, and up!!!

The last section of track to the house was truly beautiful along the top of a ridge with high picket fences either side and behind them traditional tukuls - round, thatched roof huts. Apparently, all the houses in the area are owned by Dr Fikru's relatives. He is one of eighteen children. His father had only one wife for the first four children, then she died, and he took on another two wives.

We walked the last section of the path being greeted by many locals and arrived at the compound of Fikru's father. The compound has an old rectangular mud hut with a corrugated iron roof and a number of tukuls, some used for sleeping and others for cooking and living. His father was in the compound and still quite active at eighty-two, but with a little bit of dementia. The uncle is ninety and has large thick round glasses and he came walking down the track from his hut. They were both delighted to have us with them and very pleased to show us the progress on the new "house" which was to be a large rectangular building with timber and mud walls and corrugated iron roof. The floor will be dirt.

When we arrived (at about 10 am) there was six men working on the house including one man who is the main carpenter and carried a few tools about. Slowly the numbers in the compound swelled until there were dozens of children, from infants to young adults, most of Dr Fikru's brothers, and all the sisters, wives, mothers and female cousins (who were all busy in the kitchen area) and plenty of other people who were presumably relatives

Dr Fikru was obviously extremely popular with his family and a complete dynamo, constantly whipping all the workers into action with much friendship and hilarity. He kept sitting us on stools with his father and uncle and any other oldies that appeared from time to time, but Pip and I kept slipping away to help dig with crowbars and carry logs and it gave us the chance to chat with some of the younger ones who spoke a bit of English.

Soon after we had arrived, he took us into the false banana plantation and explained how it is cultivated and what an incredible plant it is. It is obviously especially important in the area as there were many plots of a hectare or more all over the hillsides. Apparently, it is drought tolerant and so during the devastating Ethiopian drought of the early 1980's tens of thousands of people were brought from the north and east of the country to the highlands because this food was available.

It looks like a banana tree with a really thick stem up to 70centimetres in diameter. It grows from cuttings in about three years, and then the thick stem is chopped up and buried in the ground for six months. The huge leaves are sold for wrapping up items to cook and often we see donkeys laden with them going down our road to the Monday market. After six months of fermenting under the ground the 'ensat' is dug up, dried and pulverized into flour for local "bread" making. It is very good nutritionally and can form the basis for a healthy diet.

After our inspection of the plantation, we were taken into the old hut, which was very dark, with a couple of mattresses on the floor and walls covered with English language newspapers from around the world. About ten of the workmen joined us in the hut. Dr Fikru produced a bottle of purchased water that he had brought from Addis (presumably for himself) and insisted that we have it. Then one of the ladies brought in a huge loaf of bread, or 'dobo' made from the false banana flour.

The loaf was at least 70 centimetres in diameter and very thick. Traditionally, the father would cut it and so he poked the knife in the first time, then Fikru took over and cut it into huge chunks for us all.

Some non-alcoholic roasted barley drink was produced and here I think I came unstuck as it would have been made with local water. Pip wisely did not partake. Anyway, we consumed our bread and something that looked like roasted sago but was made with ensat. The women brought in little cups of hot sweet coffee very quickly. Then the next adventure was to begin.

Fortunately, we had the camera which we used heaps throughout the day as nobody seemed to mind and by now we felt very much part of the family.

As we walked to the farm earlier, we had been shown where some men were preparing a huge log to be used as the main beam for the roof of the new house. Now armed with four carrying poles and some old rope Pip and I headed with thirteen men about a kilometre up the track and part of the way down the other side of the mountain into the eucalypt forest. The side of the mountain was so steep the only way to stand easily was to find an old goat track or some other ledge. Four poles were slid under the beam and roped on. The beam was eight meters long and had been squared off to about two hundred millimetres. It was very green eucalypt and must surely have weighed close to half a ton. With lots of men on the poles and me on the front-end scrambling for a foothold (and Pip armed with the camera) we extracted the log from the forest and started to carry it up the hillside.

Helping carry a log out of the forest for housebuilding.

As always, everyone in the team (except me) had an opinion as to which way to go and how. Suddenly it was hoisted onto shoulders, and I was not sure if I was lifting or swinging at that point. Up we went on this incredible slope. We set off with enthusiasm and a couple of hundred meters up to the track we got into a pattern of getting one's shoulder under a pole as one of the others looked to be flagging, then after a minute or so changing over again, like relay runners.

I tried to stay at the back as I worked out that as we went over a hump in the track the guys in the middle took most the weight which was not good for my back. The local men are so incredibly strong and were in good spirits all the way. I overdid it a bit as I sure had a sweat up when we got back and deposited the log on the grass outside the gate into the father's compound.

More men got to work with axes to finalize the shape. Meanwhile others were putting up split eucalypt logs onto the walls of the hut which will be covered in mud during the wet season. Meanwhile Pip snapped away with the camera and Fikru kept all the troops happy. At that point – mid morning - an old water bottle filled with Araki (I reckon it was pure alcohol) was produced with lots of small glasses. Everyone stood about and had a glass and of course I "had" to be a part of that. It is vile stuff; way stronger than any spirit I have ever tasted. Because Pip and I were such a novelty someone was always staring at us and although I managed to (deliberately) spill a bit, I drank most of it (the second undoing for the day for me I suspect).

Suddenly Fikru revved everyone up, and the building was back in full swing. He kept telling us that if he left, then the job would slowly grind to a halt, so he was keen to get as much done as possible whilst he was there. Time came to move the beam in through the gate and up onto the house. Pip fortunately had the video running and got an amazing bit of footage with all the yelling and shouting as about twenty men all had their say at once. I stood back as I could see half a dozen men quite likely getting squashed at any moment.

Somehow, they manoeuvred this huge log up on top of the house end walls and fixed it in place ready to carry all the roof. At one time I swear one of the workers had the whole weigh of the thicker end on his own, lifting it up the last bit up. I pointed out to Fikru that I would like to take that man back to Australia to build me a house!! Fikru translated to all and sundry and everything ground to a halt for a few minutes until all the laughter had subsided. He was a beautiful looking boy with large black muscles popping out of his tea shirt, like so many others in this country.

At about 1.00pm three legged stools and planks started to appear and form a circle in the compound. Fikru told us that lunch was to be served but the ladies had run out of water, so they had headed off with the pack donkeys and twenty-five litre plastic drums to the nearest river. Turns out it is a couple of kilometres and a thousand feet down to the river from the village.

Fikru took us for a short walk to a spot where we could clearly see the valley into which the women descend when they head for water. It looked so, so far away and a long way down. Anyway, they were back in half an hour, and work stopped. By now there were thirty people sitting around the circle and the women were flat out carrying out huge plates of injera and bowls of wot. They go around first with a jug of water and a bowl and everyone rinses their hands as of course the meal will be eaten with fingers. We were each given a plastic plate. Most the men took two injera, but Pip and I only took one as that is plenty for us.

Sharing lunch in the highlands.

Of course, I was not allowed to have just one, so a second one was put on my plate (third mistake to the day). Then bowls of wot were brought around and spooned onto the injera as each person unrolls it on their plate. The beef and potato were a bit spicy (fourth mistake), the beetroot mix delicious and the pea or lentil (sloppy stew called 'shiro') also good. Tella – the local beer was then brought around (fifth mistake and by this stage I knew that I would regret tomorrow but hey, where else can you get this experience?). Finally, when I was sure I would burst, another glass of the vile araki was handed around. Again, I tried to spill as much as possible but had to drink some (sixth mistake).

It was now close to 3pm, and Pip and I were wondering about getting back. We had assured Fikru (without great confidence) that we could walk back home but he had said earlier that he would take us back. Anyway, he decided that he needed to stay and keep things progressing. His father tried to insist we stay, but we said no, so Fikru then conscripted one of his younger brothers (a truly gorgeous young man who was in year nine at Guder school and who was keen to practice a bit of English) to be our guide back to Mutulu.

We bid our farewells to all and sundry and headed straight up a path about a kilometre, whilst also climbing a few hundred meters of altitude to the very top on the mountain where there is a lovely new round concrete church only two years old. Obviously Fikru had been the driving force behind its construction. More wonderful photo opportunities because we had three hundred- and sixty-degree views of the Ethiopian highlands to the south and west and lowlands to the north and east. We could see Mutulu in the far distance and further down to Guder. (Mistake seven was starting to dawn on me as it looked many, many kilometres back to the farm.)

Our young guide headed off the other side of the mountain like a young goat, and we followed along slipping and sliding down the mountainside. Down, down over ridges and gullies and one stop to ask directions, first yelled to a girl shepherd across the valley then from two men who also answered the call. Our guide was not sure where to cross one of the deep gullies which can change with each wet season.

As the country flattened out, we got to the road that we had driven up earlier in the day. We walked down it for a couple of kilometres and then took a short cut heading for Mutulu. On that track we met three of our guide's relatives who were walking up to the house building project. What a slog that would be – going up the mountain!! We had a chat, and they seemed thrilled to meet us and then we pressed on to Mutulu – seeing a few monkeys as we walked.

At Mutulu, we bid our guide farewell and he said he had to buy some sugar and then head back. He must be so fit. We are not sure if he took the road or went back over the mountain. We stopped for a cold drink at the store in the compound of the sawmill operator who we had met previously. He saw us and shouted us the drink, but my tummy was so bloated and uncomfortable I could only take a few sips before the spirit I had drunk hours before was regurgitating back up my throat.

We then set off on the last seven kilometres down the road to the farm with legs starting to feel very jelly-like. All the way we were accompanied by children and the inevitable shaking the hands of adults walking towards us. Tibebu rang at one point worried as to where we were because the guard apparently did not understand when we said we were going to Mutulu, and we certainly did not expect to be away until 6.00pm. Good to know they are looking out for us.

Pip as always was powering along but I was knackered. Certainly, no dinner required but a nice cold shower to wash off the sweat and into bed at 8.30. Up and down a bit in the night, swearing I would never touch tella or araki again.

Chapter 4

May 2014 - Dealing with Unrest

EE.10: Safe in the heart of the community

Our personal security was obviously something that our AVI manager Demiss in Addis Ababa and everyone further up the line, all the way to Canberra, were constantly worrying about. Pip and I never dwelled on what may eventuate, and we lived for the moment. At all times we felt we were in the embrace of a loving and caring community who would do all in their power to protect us. We were also well aware that in the Ethiopian culture, conflicts were between local people and should never involve foreigners ("farangi" as we were called by all and sundry). I can honestly say that we never once felt fear.

At about this time Lechisa got married, and with his young bride, Zinash, they came to live in the dark rooms adjacent to our courtyard. How lucky were we. They are such delightful young people and so keen to learn. They had been living in Guder and Lechisa was walking up and down the seven kilometres each day. He is such a smart young man and loves learning from us – not just about agricultural issues but about management styles and techniques. About a year later, as word spread about the success of this farm where, by now he was the deputy manager, he was head hunted by a huge private company to manage a section of an eight-thousand-hectare coffee plantation in the south west of Ethiopia.

When he and Zinash moved out it was a great loss for us and the farm, but we really felt that the training, knowledge, and techniques he had picked up from us were very much part of our long-term legacy to the agricultural sector in Ethiopia. After all, our assignment was to be "farm management mentors".

Day 75

The managers went off to Ambo again today in the Hilux ute, but we stayed at the farm and gave them a shopping list including a tap for the biogas plant which we are happy to say is now regularly producing bags of biogas. We are not sure if the student demonstrations which were occurring yesterday in Ambo are still going on, but we thought it best to keep well away from town for now.

During a farm walk, Pip and I planned the future compost/dung pile design. Every day, two men carry the dung on their pole and corrugated iron "stretcher" from the cow shed and the cow yard and dump it behind the shed. Now it is dropped randomly by the cow-boys wherever they think is easiest. This makes it hard to pick up in the future but with a bit of organization we can have it dumped so that when it has composted, it can be easily picked up in the tractor and trailer and carted to the paddock and used as fertilizer.

Management got back late in the day with quite a story of adventure. They got into Ambo and managed to buy our bananas and oranges and phone cards - which we need to get each month so we can recharge our phones for calls and using the internet. Things were very tense with demonstrators, army, and police about. At some point shooting started and so they ran back to the vehicle and drove to the bread factory on the edge of town and sheltered there for over an hour until they thought it safe to come back to the farm.

They have all been on their phones most the afternoon as most of them have family living in Ambo. They tell me that two students have been shot. No confirmation, although when I later checked the internet news reports are saying that two students were injured by police shooting yesterday so maybe it was worse today. We are thankful we were well out of the way, and we feel perfectly safe. Demiss has been in contact with Tibebu, and we are not to go anywhere until advised and to be ready for immediate evacuation.

Day 76

I guess we should have expected the day to bring new twists and turns and sure enough it has. When we got out of bed there were very few of the managers about and those who were here were constantly on their phones. Turns out that a boy down the road had hung himself yesterday. Due to the roads all blocked in Guder and all the police otherwise engaged, the local heavies including Tibebu decided that the family could bury him today without any inquiry/ inquest. Hence most people were tied up for the day with yet another funeral.

Meanwhile things were developing with the student demonstrations in Ambo and Guder. With over 9000 students in the University campus and many more thousand high school students in the local towns, things can get out of hand very quickly. We had phone calls from Demiss and a guy from the embassy who were both worried about us. The embassy said that security people were about to leave Australia to come and assess our situation. Apparently, the embassy has hosed that down for now!!!.

We kept emphasizing that we felt perfectly safe as seven kilometres away seems a good distance from our perceptive, but I guess it looks pretty nearby if you are on google maps in Canberra. They were mostly worried that if we have a medical emergency, they would not be able to get to us.

Tibebu and others also confirmed that a boy had been shot by a guard in Ambo yesterday as he fled from the police into the main hotel (which ironically is meant to be our "safe house" if there is trouble in the area!!) As a consequence, the demonstrators had attacked the hotel smashing up the ground floor and burning a number of cars in the hotel compound. (Quite a big issue to our mind as that would be about half the cars in Western Ethiopia by our reckoning.)

Our American friends in Guder also rang to tell us that Peace Corps was evacuating them (and their colleague, Josh, who had shooting outside his house in Ambo) to Addis. We spoke to them later in the day and they had arrived safely in Addis but spent some time sheltering in the police station at Ambo after their trip from Guder.

We had a walk around with Lechisa showing him some of the changes around the farm since he headed off on leave nearly two weeks ago for the mourning of his brother. He is pretty happy again back at work but is wearing a black baseball cap as his sign of mourning. It actually makes him look like a movie star but for some reason when wearing it he looks so black you would only see a silhouette in a photo.

Late in the day we took Tibebu down to the compost area and discussed with him our ideas for laying out dung from the dairy sheds so it is convenient for the staff to dump as they clean the sheds, but also convenient for the tractor and trailer when it is composted in a year or two and can be carried out to the fields as fertilizer.

It is a typical example of what I think our job has become to some extent. The management seem to be able to manage today but cannot think much into the future, so we are trying to set things up so there is a "recipe" for everything which incorporates some consideration of the future.

Tibebu and Lechisa were both thrilled with the idea of dumping it somewhere it will be easy to access. Also, due to the language barrier if we just show them our idea, we can never be sure that they understand (and can be pretty sure that the wrong message will get to the cowshed staff) so we made up white pegs and had it all pegged out for them.

We had a long walk with Tibebu up the far end of the farm to where the oxen have been preparing some land for wheat. It's a very rough area being cultivated and even Tibebu marvels at the job the oxen can do.

Up in the corner of the farm, we spotted a local farmer grazing a dozen animals in a bit of native forest on our land. Pip and I reckon it happens every day. Tibebu then heard an axe chopping in the farm forest area, so Pip and I froze, and Tibebu tip toed off into the forest. Meanwhile a lady appeared close to Pip and I, heading off with her oxen under the fence where she had previously propped it up to let the animals in. Suddenly there was a yell, and a young lad was running for the forest. Tibebu appeared with a huge grin and an axe over his shoulder. He had crept right up to the intruder in the bush and yelled "drop it". The young fellow got the fright of his life, dropped his axe, and ran for his life.

As we got back to the courtyard Tibebu could not wait to tell all and sundry about his escapade and show off the axe. (I suspect it will get returned after a while, as it would be the equivalent to keeping someone's car in Australia.).

Tonight, we had a call from Demiss telling us that we are to be evacuated tomorrow if possible. A demonstrator was shot and killed in Guder today so there is no telling what will escalate or for how long. Demiss will be ringing the police in Ambo and Guder tomorrow morning and if he can get a vehicle through, he will come from Addis and pick us up.

We are to be ready to go at a moment's notice as he does not know when it will be or how many days we will be away. It is difficult to know what to do with food in the fridge etc. and how much stuff to take. Pip is happy to be off, but I am reluctant because 1) I think we are much more vulnerable on the road than here safely at the farm and 2) once they get us away, Canberra will be really reluctant to let us get back, given how worried they were before we ever came.

Presumably tomorrow we will be writing the blog in the Kaleb Hotel in Addis, but who knows what the day will bring.

EE.11: Evacuation

Demonstrations can turn violent very easily in Ethiopia and Ambo is one of the most difficult areas for security in the country. There is a huge population of university and high school students ready to ignite at any time and if they can mass on the bridge in Ambo, they can cut off the western part of Ethiopia. Traditionally battles are often fought between police, military, and demonstrators for control of the Ambo bridge.

Day 77

Well, it turned out to be quite an exciting day, but we are safely in Addis. Up in good time and packed up whatever we thought we may need for the next few days - always with the thought in the back of our mind that we may never be back. The phone network and the power were still working so we had calls from the embassy to say things had been really bad in Guder and Ambo yesterday. Officially 11 (later revised to 19) young demonstrators were killed by police and there was a lot of damage.

We managed to get on to Demiss and found out that he was coming to get us and had spoken with the police chiefs in Ambo and Guder and been told that he could get through as long as it was before lunch as the funerals would start after lunch and more trouble was expected then. Apparently, the police and army need the power and phones, so they have not shut down the networks.

Unfortunately, due to the heavy rain overnight our road was only just passable for a two-wheel drive van so when he arrived poor Demiss was very tense. We said quick and fond farewells to the farm staff, gave away the perishables out of the fridge, and jumped into the van with our backpacks and computer bag. We slithered and bumped our way down the road to Guder with Demiss struggling to keep his van on the road.

On the outskirts of Guder, we started to come across half dismantled barricades and roadblocks of huge rocks and logs. Also, there was a big tent being set up on the road – presumably for the funerals later. There was none of the usual hustle and bustle and none of the usual donkeys and other animals. Everything very tense and heavily armed soldiers everywhere. We did not notice much physical damage in Guder but very few vehicles about and certainly no cars other than us.

Another tent was being set on the road as we left the town as well. A sense of relief as the first hurdle was over and the next hurdle was the twelve kilometres to Ambo. On the outskirts of Ambo, we were stopped by a police roadblock and thought that may be the end of our evacuation plans but after some talk, three officers got in the van with us, and we headed into town.

As we got close to the university there was again many partially dismantled roadblocks and barricades of huge rocks and other debris everywhere. In the main street we let out the police officers and picked up an elderly British lady who is working at the Ambo University for Volunteer Service Overseas (VSO), a British NGO. VSO had no way to evacuate her, so had asked AVI to help. Interestingly, yesterday she was with the Peace Corps people as they were being evacuated and even though they had room in their car and the local Peace Corps volunteers wanted to take her, the management of Peace Corps said no. (Some around here say typical of the US, but we do not know enough to judge). She was left stranded in Ambo until we picked her up. She was very shaken as there had been bodies in the street outside her gate and substantial amounts of shooting for the proceeding couple of days.

As we drove through the main street, it became clear that this was truly a major disaster for the town (and presumably for the country in terms of ethnic unrest). There were piles of tyres and barricades (still smouldering), two upturned semi-trailers, other trailers completely burnt out right in the middle of the main street, a couple of burnt-out four-wheel drives in the middle of the road, and one three level building burnt out completely.

The Abebech hotel (our "safe house!!") looked significantly damaged with all the windows broken on the ground floor and in the carpark at least seven Land Cruisers that were burnt out. Apparently, a guard had shot a youth who had started vandalizing the hotel, so the guard was attacked and killed and then all the vehicles attacked. We had to travel very slowly and negotiate all the debris on the road and the masses of people that were walking up and down most in their white gabis getting ready for funerals.

For the three kilometres out of Ambo towards Addis, which is in the direction of the engineering campus of the university, there had clearly been battles with the army not long before as there was debris, smoking piles of tyres on the road and destruction everywhere. A large new building close to that campus (where we had stopped for coffee a couple of weeks ago) had been completely trashed and looked so sad with every one of its hundreds of windows broken. It had been opened only a year or two ago by an enterprising young lady.

Slowly the road became clear, and we were headed for Addis although at one stage we came across hundreds of large rocks that had been rolled onto the road since Demiss had come through an hour or so before. We could just pick our way around them following a truck that was also heading for Addis.

The VSO lady talked nonstop all the way with more tragedies and rumours, including that at least twenty children had been shot dead at a high school in Ambo. Incredibly it is quite possible. From the western world, we cannot conceive the ability of the government to censure news and information. Even over the last few days there had been virtually nothing on the radio or TV in Ethiopia and there is nothing in the newspapers here in Addis. Also, they manage to keep most the websites down so there are only snippets online. It would be far too dangerous to take photos of the destruction in Ambo – tempting though it was.

We had an uneventful drive into Addis where we dropped off the VSO lady and then at 3.00pm, we found our way into a café for our first food or drink since our morning coffee at the farm. There we were met by a guy from the Australian Embassy for a good long debrief. We made it clear we are keen to get back to the farm ASAP as it is still very safe there in our opinion.

We are told there is no chance of travelling anywhere before next week as simultaneously to the problems with the Oromia people demonstrations, there is to be a big opposition party rally in Addis on Sunday, so we are meant to lie low. He also told us of a bomb that had exploded in a cafeteria of a regional university where students had been watching a soccer game on TV. It had killed about twenty of them and injured many more. The embassy reported that they were not sure who was responsible but that it could mean more violence.

We returned to our friendly environment at the Kaleb Hotel where, as always, we were given a great welcome by the staff. For dinner we walked up the road a short way to where a band was playing in an outdoor restaurant. Great African music accompanied by a pizza and a glass of beer. It seemed an awful long way from where we had started the day!!

Not sure if it was the rain that kept the beggars away or if there has been a clean-up since we were here a month ago but there were no beggars as we walked in the street which was a relief!! The staff here at the hotel think it may have been a clean-up for the visit a few days ago, of Senator John Kerry, Secretary of State from USA, who has been here discussing human rights issues in Ethiopia. Timely eh!!

Chapter 5

May 2014 - Life Goes On

EE.12: Public Transport – Addis Ababa

Ethiopia has a public transport system that operates on several levels. In the countryside, up to about 150 kilometres radius of Addis Ababa, there are numerous white Toyota vans that run every few minutes into the city. For us it was possible to use these vans from their regional centre of Ambo into and out of Addis Ababa for a very small cost, but they did not extend out the next 12 kilometres to Guder. The vans only depart when they are full but, with hundreds of people at the crowded bus stops and numerous vans ready to go, there are departures every few minutes. If a passenger gets out at a village along the way, then the bus will pick up an extra person along the way – ensuring that the trip is always crowded and uncomfortable.

For areas a further distance for Addis Ababa, there are two bus lines (Sky and Salem) that operate large 40-seater buses that mostly run to a timetable. We had just one journey on a Salem bus from the southern city of Hawassa to Addis Ababa – a journey better forgotten - and made a resolution that if ever to be repeated, one should not sit at the front of the bus where one can anticipate scores of likely disasters ever few minutes!!

In Addis Ababa there is a bus service which is mostly used for transporting workers to and from their workplaces. There is also a newly completed overhead light rail system with two lines – one running north -south and the other east – west. Then there are the relatively expensive yellow airport taxis, some of which have a meter but mostly operate by negotiation. We found the cheaper blue contract taxis a very convenient means of getting around the city. The vehicles are very old and dilapidated, and negotiation is essential to agree at a price, but *they could get us across the city for 200 birr ($10) in relative comfort. Finally, by far the largest proportion of local people use the line taxis – or blue donkeys as they are known. These are inevitably very old Toyota HiAce vans painted blue, and it is possible to travel to most areas of the city for about 5 birr (25 cents) – so a very cheap form of transport – but not easy or comfortable for the uninitiated.*

Day 82

Today, while we are staying in Addis Ababa, we decided it was time we became comfortable with using the line taxis to get around Addis.

It is one of the most amazing laissez-faire systems that could ever be designed and incredibly efficient for moving millions of people about. This is how it works.

Literally thousands of these very old and battered mini vans run between points of the city. They do not necessarily take exactly the same route depending on traffic or roadworks and if there is not enough patronage on one route, they simply change to another at the discretion of the driver and his offsider.

There is nothing formal about the location of the bus station, except one at Merkato which is where all the buses from the countryside come in. You must just "know" this - like so many other things in Ethiopia. And you also must just "know" where the buses at that stop are likely to be going. There may be five or six destinations from the stop down under the bridge at Bole, but if you want to go to a stop different to what is offered, you need to "know" where another bus station is that may have buses going to your area of interest.

When you get to the bus station (mostly just the side of the road somewhere but occasionally at a large car park), there can be anywhere from ten to two hundred blue vans waiting, with people coming and going in all directions. Each one has an "offsider" hanging out of the window or walking around yelling out the destination of that van.

They always yell in Amharic and somehow you just must "know" that they use Amharic slang or nick names for different spots - words that appear to have no relationship to the destination that you think you want. Given we have no idea about the "slang" naming, we walk around saying for example 'Mexico' for Mexico square, or maybe 'Bole' for the area of our hotel. One of the guys from the vans will then either stuff you into his van or wave you in the direction of another boy who will be the offsider to another van.

When they have the van full, or more often way overloaded, depending on whether or not they have heard if there are traffic police around that day, the offsider jumps in and off you go. The offsider then pushes and shoves around in the van and money is handed around until everyone has paid the three birr or so. From then on you can call out "worage" at any time, which means 'stop' and you can get out - or you can stay in until the bus gets to its destination, and everyone gets out.

If anyone gets out enroute, the offsider then hangs out the window and continues to yell out the destination of the bus. The bus will then stop anywhere to pick up extras until it is full again and the offsider goes quiet. When the bus is empty at its destination, it does a U turn and joins the cluster of buses going the other way and starts yelling again.

If there are too many buses at that stop going to one particular place but not enough commuters wanting to go there, they just start yelling a different destination and so the system keeps providing the service to where the demand is greatest. Miraculous. Somehow it works brilliantly and is so cheap.

It's just a matter of finding the right bus!!!

Yesterday we thought we would go and visit the American Peace Corps couple Terry and Ellen who have also been evacuated from Guder and are in a hotel in a different area of the city. They had told us to walk to the Bole stop and get a line bus to Mexico square and then we were to get another bus to Mekanisa. Everything went well for the first stage. We let off a few passengers and rounded up some more until we were full again to head on to Mexico.

We arrived at Mexico and we got out of that van, but we had no way of knowing where to find the vans that would be going on to Mekanisa. A kind lady with some English stopped and asked to help us and told us to walk around a few corners to where we found about thirty blue vans in a carpark. We kept saying Mekanisa and were ushered into a van. We waited a few minutes until it was full and off we went. Quite a long trip and eventually it ended at a large roundabout with heaps of road works. Traffic and people everywhere and we walked off in the direction that we thought were the directions Terry and Ellen had given us. We asked a couple of people, and they waved us on.

Eventually we realized that we had walked the wrong way so went back to the bus stop. We rang Terry and Ellen who did not recognize what we were describing. Eventually we worked out that we were completely on the wrong side of Addis Ababa. We had been put into a van to Megenagna, not Mekanisa. More negotiating to find the correct van to take us across the city and after a few more hassles and a final contract taxi ride to find their hotel we eventually got to meet Terry and Ellen at 2.00pm . We had a merry lunch with them – as always swapping stories of adventures and experiences, and we then walked with them along a river and through the streets to a bus stop. They are experts at the system and put us into a van that took us almost all the way back to Bole and so to our hotel. Very simple really.

EE.13: A visit to the farm by Senior Managers:

Pip and I spent ten days in Addis Ababa for the period of evacuation. We did manage to do quite a bit of research including a visit to the Salem Children's Village farm where we saw a well-run dairy and chicken enterprise and we attended the Africa Agricultural Expo which happened to be on at the Millennium Hall close to the Kaleb hotel.

When we returned to the farm, the local towns and villages were still tense with political unrest but the loving and generous reception for our return from the people at the farm was very reassuring.

Day 91

This morning when I heard the tractor start up, I knew that Merka would he heading off for some ploughing. Yesterday I had checked that paddock and cleaned up some rocks but there were half a dozen rocks that were too large for me to carry but would do some serious damage if (when) Merka tried to plough right over them.

I headed out to the yard as he was preparing the tractor and had an animated conversation trying to convince him that we should go first in the ute and pick up the rocks. Eventually I conscripted the help of a few others using hand and eye communication, and they helped me convince Merka that he should give me the keys to the ute, and I would take some men down to pick up the rocks.

As the tyres on the ute are completely bald and the 4wd does not work, it was a very slippery ride and we only just made it to the paddock, but mission was accomplished, and everyone agreed that the paddock would now be much easier to plough for the next thirty years!!! – and with no damage to the tractor or plough.

It was then time for Pip and me to meet with the senior managers of the organization who had travelled out to the farm from Addis. We said our "hellos" and then waited for them to finish the discussion they were having with Tibebu – in Amharic of course.

Managers meeting in our courtyard. The head office managers are in the two right of Pip. The rest are all our farm managers!!! Pip with the hat.

We were called into the office, and we gave them a presentation on what we had been up to for our first three months and what plans we had for the future. We got a good hearing and lots of affirmative comment and I think they realized what an excellent relationship we have with Tibebu. It was a good chance for us to further press for some of the most important changes that we have not been able to implement. Luckily, the heavies from head office told us that they had told Tibebu that he must pursue our suggestions.

It also gave us a chance to introduce our concept of setting up a replacement fund that must have some money put into it each year, ready to replace assets and machinery so that the farm is a long-term business and does not rely on intermittent charity donations for its future. This is something we are determined to have in place before we leave. I can't say the heavies were exuding enthusiasm for the idea! Same old story – Ethiopians live for today and cannot think for tomorrow. We spend a lot of time trying to change that attitude on the farm – farming is, really, all about planning and long-term thinking.

The two General Managers from Addis are very nice old guys and we had met them a couple of times in head office. Presumably, the guy who is Deputy Director of the whole organization will become Head when dear old Abebech Gobena finishes up in the not-too-distant future. They hope she may make one last visit to the farm in a month or so. I think she will be proud of the farm to see it all cleaned up and looking productive.

They insisted that we join them for lunch in Guder and we were about to head off when Getun, the dairy manager, appeared at our hut with a huge thorn embedded in his thumb knuckle. Pip and I got out with our headlights (like surgeons), needles, scalpel, tweezers, filtered water and bandages. Much to the amazement of the head office guys, we went to work on the thorn whilst they and our farm manager, driver and deputy manager all had to wait. Little did they know that barely a day went past where we did not treat someone from the community for illness or injury.

With my best surgical look - operating to remove a thorn from Getun's knuckle.

We managed to remove the thorn and dressed it with antiseptic and strapping and told Getun to come back tomorrow – quite expecting that we may not have got it all.

We got to Guder and were surprised to find hundreds of students milling about, some huge army trucks parked and heavily armed soldiers around the school and other areas of town. Pip and I were pretty suspicious that something was up as our two vehicles did not stop in Guder as planned but went on to a restaurant in Ambo without explanation.

We had lunch at the old Hailee Selassie era Ambo Hotel and from there we snapped a photo of one of the burnt-out trucks that is still in the street. We bought bananas and a couple of electric fittings and headed home. The army was still around in Guder but no students.

As we later found out from Zinash (the wife of Lechisa) who is still finishing year twelve at the Guder school, during today the year twelve students were about to sit an exam when some army guys came to the school and arrested one of their classmates. This resulted in about ten students with rocks taking the other twelve hundred students' "hostage" - forcing them to a sit on the oval as a show of force to the army so it would allow the arrested boy to sit the exam. From the description from Zinash who had been one of those sitting on the oval, it was all very tense until a "religious figure" came to the school and negotiated peace - fortunately before the firing (slaughter) started again like last week.

It seems incredible that five students were confirmed shot dead by the army in that school last week and at least another fourteen university and school students confirmed dead in Ambo. How would you like to be sitting an exam which determines your university placement, what course you will do, and future of your life after a week or two of serious unrest!!!

Presumably everything will settle down soon as the students start two months' holidays any time now.

EE.14: Frustrations

Back at the farm but not excited about progress – or lack of it – whilst we had been in Addis. We could probably fill a book with all the times we felt frustrated during our two year's stay, but mostly we put them behind us quickly and set out to resolve the next issue rather than constantly dwelling on the negatives. It was never about improving our lives, but always about not being able to help the local people of Birbirsa, the AGOHELMA organization to which we were attached, the agricultural industry of Ethiopia or the country as a whole. Sometimes we had to write it down to ease the burden of frustration.........

Day 94

I felt frustrated today with all the inefficiencies around and felt like heading down to the sheds where all the slackest managers hide and giving them a full-on blast. I could fill the next three pages with the annoyances that I have seen today but suffice to say the general manager and deputy did not show up, the only lady who has a key to the milk processing room did not show up until nearly nine, so I waited at the milk room for nearly an hour - along with the two milking ladies and ten thousand flies getting into the four milk buckets that do not have lids. The milking ladies keep the flies out while they are in the milking shed but once the cow-boys carry the buckets full of milk to the concrete outside the locked door of the processing room, apparently it is no longer anyone's responsibility, so they all sit on the ground waiting while the flies help themselves. AHHHH!!!

Behind the calf shed the labourers are shovelling compost into bags, putting it on the backs of three women who were then carrying it half a kilometre to spread around the newly emerging maize crop. Turns out the driver Merka should have been driving the tractor and trailer to cart the compost, but he had decided to take the ute to Ambo for no apparent reason, and he is the only one with a key for the tractor.

If I had the key, I could drive the tractor and make the composting job so much more efficient. He had also taken Abissa, the manager of stores, can you believe - with him (we call him Mr Keys, as all he ever does is carry around a large set of keys). Strangely whenever we want to open a door to get a tool or sprayer or whatever, he apparently does not have the keys for that door!!!!!!!. At least the grain grinding manager was grinding a bit of grain for some local women late this afternoon. This is a small contract business operated by the farm to assist local people.

Pip continued her pruning of some long-abandoned apple trees out behind our courtyard, and I did some spraying with Roundup with my little one litre hand pumper. I then spent ages on the 'internet seat' in the paddock as our Gmail offline App was not working properly and we had some important emails to send. We draft the emails back at the hut "offline" where we do not need to hold onto an umbrella at the same time as the computer to keep the glare of the sun off the screen.

I took Lechisa to the cabbage seedling plot to do some random counts to get an estimate of the seedling numbers. They had been saying how they needed to plant 40,000 or more seedlings but our counts showed there is only 35,000 growing in the plot. I suggested they could prepare more plot and plant more. It will be interesting to see what happens.

Despite the fact that there are paddocks which desperately need cultivating before it rains heavily again, tomorrow is one of the "Saints" days for which Ethiopia is famous. It is sort of an optional public holiday, so nobody really knows who is "working" and who is not, depending on the strength of your religion. Apparently, there are dozens of Saints days throughout the year, and the management declare that as part of the religious observance of the day, there can be no cultivation done (really??), but the tractor can pull the trailer. I suppose the senior managers will stay away again tomorrow.

Day 95

Numerous frustrations from yesterday forgotten after a good night's sleep and we managed to kick some goals today. Tried a few more adjustments on the new cream separator and got it running almost perfectly by afternoon milking. The amount of cream and butter has gone up enormously since we replaced the old separator, so everyone is delighted, and the operation is only taking half an hour instead of about 4 hours. A little of the milk from the dairy is sold as fresh milk – especially to us – but most is put through the separator. The cream is then made into butter to be sold locally – or taken into Addis Ababa for selling to the staff and community there. The "skim milk" is then made into cottage cheese here at the farm and again sold locally or in Addis Ababa.

I managed to find Mr Keys and get one of the biogas bags that we had heard were deep in the back of a dark storeroom. It is a large, white, closely woven poly bag with a built-in valve/tap, about two metres by one and a half metres. Pip and I attached it to the biogas plant, and it has been filling up nicely all day. Hopefully by tomorrow the heat for the cheese making will be from a biogas fuel source.

Carting a bag of biogas to our hut with Lechisa and young Lami.

Most of the farm staff and some managers were today working on building a new "post and rail" fence for another calf paddock. So much for the holiday – we never really know what to expect and I suspect it is likewise with the managers. A lot of the staff are day labourers, so I suspect they show up if they need the extra wages.

Pip and I have been pushing the point for ages that we need to rotationally graze the young calves but there was only one small paddock (half acre at most) that they were put in every day. By tonight the next paddock is nearly finished so we can tick that off our list. Also, Tibebu showed up late in the day and was in a very "listening" mood – which is always good for us.

Fortunately, a cow had just calved. It was one that we insisted be "dried off" three weeks ago that otherwise would have been being milked from last year's lactation right up to the next birth which greatly reduces the productivity of the following lactation. We were in the right place at the right time with Tibebu and we managed to get her put in a separate shed with her calf instead of being tied up in her usual place in the dairy shed with her calf in a different shed. With our ongoing pressure Tibebu is also closer to ensuring that the calves get enough milk to grow properly for the first three months but that is not bedded down yet.

To try and give some incentive for the managers and guards to get off their bums and do something productive for themselves and the farm, Pip and I came up with a scheme which has been endorsed by Tibebu where anyone employed at the farm can claim an area up to 500 square meters and can grow whatever they like. There will be a prize at the end of the year for the most productive plot and also the worker can keep or sell the produce for themselves.

There are now already eleven plots that been started so the enthusiasm for action is certainly still alive. They range from a few square meters to a few hundred square meters. Our personal tools have been used extensively, and the best thinkers have been asking us for seeds and seedlings. Almost all the farm workers walk past Pip's gardens every day and see lots of vegetables (sweetcorn, lettuces, beans, cucumber, peas, and tomatoes) almost at picking stage and we encourage them to think they can do the same.

It is immensely satisfying seeing their plots develop and is beneficial for the farm as it develops enthusiasm and we watch and learn the lessons of successes and failures in the local conditions to see what may be possible on a farm-wide scale. One young, strong, and enthusiastic guard, Carson, is even successfully growing tobacco on his plot, but we are not sure about encouraging that. Carson is a very hard worker and is one guard that helps with a few other jobs around the farm. He also has developed a very productive sugar cane patch on his plot.

Pip did some more apple tree pruning and I gave her a hand, sawing off some big branches. In the evening, we lit the campfire and because Merka and one of the guards were about, I invited them to join me for a glass of Tej. This is alcohol made from honey (a bit like mead but not so sickly sweet) and was given to me by the head office when we were there a couple of months ago. We had a merry half an hour of pre-dinner drinks as they have enough English for a conversation.

EE.15: Security and Safety:

Whilst Pip and I toiled away at the farm, others were continuing to worry about our security. After discussions with the Embassy, Demiss had concluded that in order to continue with our assignment, we had to have an escape route. Due to the nature of the landscape in western Ethiopia there is only one main road to the west. Whoever controls the one bridge in Ambo controls travel into or out of the west of the country.

Unfortunately, our farm was a little past the bridge, so if it was taken by demonstrators – or the police or army - then we could not get to Addis Ababa. None of the locals seemed to be sure but Tibebu was of the opinion that the road which travels up past our farm continues for a very long way –eventually over the mountains and down to the main road that runs from Addis to the southwest of the country. Could this be used as our escape route if required – or at the very least could we make it sound like it could be used, so that Canberra and the embassy would have confidence about our ability to escape?

Day 101

We suspected a huge day and certainly it lived up to expectations. It had been organized that our country manager, Demiss, would come out to the farm and we would go in the farm ute with our farm driver Merka and farm manager Tibebu, to investigate a back route out from the farm to Addis Ababa. This idea makes AVI and the Australian Embassy much happier about our posting as they are still worried that there will be more trouble in the towns on the main road back to Addis, including Ambo and Guder.

We had ensured the ute was full of fuel but were worried about the bald tyres and the likelihood of a break-down and had warned Demiss of the dangers!! He arrived here about 10.00am with Tibebu who had been in Addis for the weekend. Pip and I had packed up a few supplies (including our sleeping sheets and a torch as we foresaw a very likely stay in a flea ridden hotel somewhere), plenty of water, biscuits and fruit and the usual toilet paper supply, umbrellas, and alcohol hand wash. Predictably none of the others took anything at all.

After a coffee, we headed out the gate and turned left towards the highlands. On the whole the road remained horribly rough although there were some sections where we could get drive at over 40km/hr. We flogged along for over four hours through the most beautiful countryside and African rural scenes imaginable.

Higher and higher we went until we were well over 3500 meters in altitude. We crossed bridges and drove through areas of highland plains being grazed by horses, forested hills, over-grazed hillsides, gushing rivers, incredibly remote villages with huts, shanties, bare earth streets, masses of people in most of the villages and animals everywhere. The area was populated by strong, fit looking men, stunningly beautiful women, and gorgeous happy children by the thousands – mostly thrilled with the idea of seeing a vehicle, let alone one with a couple of white people inside.

We often had to ask directions as there was no chance of any signage and at one stage, we stopped at a World Vision sponsored community centre on the edge of a village. We waited at the ute (with an ever-increasing audience of children) whilst Tibebu went in for a few minutes. We think he was probably finding out about the condition of the road ahead but also maybe sussing out somewhere safe for us to stop in the future, if required.

Soon after we came to Shenen, which maybe shows up on a map and is the end of the bus run that goes past our farm. It is serviced by incredibly tough, mid-sized, ancient Russian-built buses that hold about thirty people and usually have huge loads on the roof. By then we had then travelled about fifty kilometres from the farm and had another one hundred to go at least to get to the bitumen road that runs from Addis Ababa to Jimma in the Southwest of the country.

The county got flatter but the road no more comfortable and we were often at 20 - 30 kilometres an hour, lurching over the huge rocks from which they make the roads. A lot of the time the vehicles try to drive in the table drain which is often smoother, but the locals keep rolling large boulders into that area to stop the vehicles driving close to their huts.

At least the rocks on the roads keep them from becoming quagmires when wet and stops wheel ruts forming. Of course, the locals are either in a bus or walking, so the condition of the road is not really of importance to them. All the way along was the most amazing soils, ranging from bright red, through to pitch black and topsoil clearly metres deep. Cultivation with oxen was going on everywhere and planting of maize and wheat in full swing with women dropping fertilizer and seeds into furrows by hand. We stopped at one stage near some bushes so Pip and I could have a pee. The locals just stand or squat anywhere and pee. Whilst stopped we shared our bread rolls and fruit with the others who seemed happy for sustenance. You wonder if they thought of that a few hours ago?

Eventually we got to the main bitumen road at 2.45pm after four and a quarter hours' of being thrown around inside the Hilux. How it stays together I have no idea. The rattling, banging, and crashing inside means that it is mostly too loud to carry on a conversation, and you must watch your head does not crash into the side door. At least the ceiling is quite high so hitting your head on the roof was not a problem.

No seat belts in the back, but I did have one in the front. We stopped on the side of the bitumen to decide what to do next and I noticed that one back tyre had canvas showing right around the tyre and one of the front tyres just has a patch of canvas showing. No one else seemed particularly interested but I was not at all thrilled with the idea of a few more hours home and maybe Demiss also was thinking this was certainly not within the safety standards in the AVI manual!!! The chances of us getting injured in the next few hours in a car accident were a thousand time higher than they will ever be from stray army bullets or demonstrator's rocks!!!

It was decided that we could not go back the way we had come as it would be dark by the time we got back, so we headed towards Addis Ababa for fifty kilometres on the bitumen, with the driver half nodding off, me ready to grab the steering wheel and thinking about the tyres, and the others all asleep in the back!!

We passed through Welikite, which was quite a big town and is on the maps, and got to Weliso, also a fair-sized university town. We stopped at a café for "lunch" where Demiss, Tibebu and Merka shared a plate of injera with Kitfo - raw minced meat and other spices.

As the only thing on the menu was injera (like most cafés and restaurants), Pip and I shared one that had a bowl of finely chopped cabbage and other veggie mixes. It was supposed to be "not spicy", but I think it was well laced with wasabi and/or chili as we could only eat a little. Probably reduced the number of bad bugs going into our stomachs anyway as everything was incredibly grotty and the toilet indescribable. Pip tells me she thought it wasn't too bad as there was "only a little bit of faeces spread about" and that's in a café!! At least there was a tap for handwashing!

From Weliso, we took a left turn off the bitumen and headed back into the highlands for a climb up to the area close to Wenchi Crater where we had been only a couple of days before for a sightseeing outing from the farm. From there we could head down to Ambo and so home. The road up got progressively worse (starting from bad) and when we were on the steepest climbing section, the heavens opened, and the torrential rains began. It was very hard to see in front, but we could not stop as we already knew we probably would not be home by dark. In Ethiopia this is an absolute "no-no" as most vehicles do not have lights and there are people and animals on the road everywhere – all very black.

There were erosion gullies forming across the road in front of our eyes, due to the volume of rain making torrential streams. At one stage, Tibebu and Demiss got out in the pouring rain (using our umbrellas) to move some large rocks into a gully to form a bridge for us to cross.

Eventually we ground our way to the top of the mountains, and the rain stopped, but up that high, there was ice from hail in the drains, and it was pretty dark and foggy. At least from here it was all downhill!!

Almost last light for the day and we were rebuilding the road as it washed away in front of our eyes.

Nearing Ambo it got completely dark, and driver Merka turned on the headlights. This turned out to be one headlight that pointed at the ground about ten meters in front of the ute. A few times he turned it off, and I thought we could see more with the parking lights which did seem to work.

I thought it was impossible to see when we got into the town of Ambo, but we headed out the main road pressing on for the last leg home. It was probably our only chance to see Ambo at night with its few bright lights, but it did not last long as they all went out from power disruption as we were heading through the middle of town.

When a vehicle came from the other direction with its lights on, we could see nothing, and I was sure we would plough into some of the hundreds of people that were walking on the road (all black) and animals everywhere (mostly dark brown). There were even shepherds moving small mobs of stock down the road in the pitch dark!!!

We got though Guder somehow and the usual thumping, bumping, lurching, and rattling up our road and into the farm gate. 9.5 hours after leaving. Demiss took his van (which does have lights) back to Ambo to sleep the night in a hotel as he needed to be back in Addis by 8.30am in the morning.

Luckily, we had some stew in the fridge which we heated up for dinner and then early into bed.

On the way to the loo before bed, (which is quite a walk out behind our courtyard) Pip's spotlight torch picked up a giant mongoose about 80 centimetres tall with a metre long bushy white tail. It was great to finally see this nocturnal animal which had made the many large holes around the farm. We will now have to Google its proper name.

EE.16: Pip's Birthday:

Our first personal milestone at the farm was Pips birthday. If ever we needed reminding what a loving, caring community we lived in this was an early reminder and one that would be repeated over and over again in different forms. We felt so humble on this occasion, and many others, when people who have so little are determined to share with us what little they have – holding no grudge or animosity at our privileged position in the western world. Even though we often ate different food and obviously had access to far more resources, I think the local people really admired and respected us for living amongst them, working with them, and as far as possible, we embraced their culture and lifestyles.

Day 104

My present to Pip of a few Snickers bars purchased on a recent trip to Addis Ababa started what turned out to be a very happy and memorable birthday for Pip.

Our first morning job was some spraying that we wanted to do for a little trial we were setting up for minimum tillage crop planting. We are worried that even if it is successful, we are not really sure if we want to introduce more chemical usage into their farming practice, although it will be much better for the soil not to be cultivated every year to get rid of the weeds. I suspect there may easily develop a culture that if some chemical is good then a lot must be better. This is clearly not a good way to work out chemical concentrations and mixing formulas!!

When we arrived back from that job, we found half of our courtyard transformed into a party scene. A semicircle of chairs surrounded a backdrop of green branches, which, along with the area inside the semicircle was covered with strips of silver, pink and white paper. This paper had been rescued from the packaging of the cream separator when it was purchased. In front of the backdrop was a low table on which there were more decorations of flowers and cut reeds. We were not sure what the timing was for the party but realized we would have to wait till the farm ute got back from Ambo - and that could be anytime.

We had our usual morning coffee, and then I went to do a bit more spraying of the channels. There was still no indication of the party starting, so we had enough time to put in about twenty white posts to mark where Pip had planted the native Podocarpus trees which she had rescued from being dug up in the guava plantation. This is an area of half a hectare that was planted to guava many years ago, but Lechisa has been keen to cut it all down and use the land for other crops as there is no demand for guavas.

Finally, we were summoned to the circle. By now the table had a vase of flowers (including a few fresh mushrooms), two candles, baskets of bread and popcorn and bananas and oranges, as well as bottles of Miranda (Fanta) and water. Stuck on the brush backdrop was a large computer-generated sign –

" Birth day Celebration at AGOHELma Campuse JUNE 29/2014."

"Happy birth day to philippa"

Underneath a couple of photos of Pip with her horses in Australia that I had previously given them. It was very moving, and Pip felt quite emotional being made to feel so special in a community so, so far from home. The office/courtyard cleaning lady, Buckalech, was seated in front of another small coffee ceremony table with the coffee pot boiling on the little portable charcoal stove. We were handed a plate of bread, roasted barley, popcorn, a banana, and a slice of orange and given a bottle of Miranda (Fanta) each.

The other two bottles of Mianda were emptied into a large enamel kettle and watered down so that there would be enough to share amongst all the staff. It was a wonderful birthday party with most the farm management staff and guards present. Zinash had organized it all and many others assisted in putting up the decorations and food and coffee making. There was lots of chatter, photos and laughter, and a bit later six health extension staff who had been in the health centre rooms, joined us.

After lots of photos were taken with our camera, everyone crowded in to see a slide show on our computer of the Wenchi Crater trip . Last weekend, we had organized for Merka to drive Pip and me and Lechisa and Zinash up to Wenchi Crater and although it is only an hour from Ambo, most the staff had never been there. They enjoyed seeing photos of us all on horses riding out of the crater. The party slowly broke up by about 4pm. Thankfully the afternoon's rain had held off until much later.

Pip and I with some of the staff at the birthday party.

Having recently seen the giant mongoose on the way to the toilet at night, tonight Pip was thrilled to see a small deer (dik dik) in the torch light on her visit – fantastic. I am sure they would not survive were it not for the bit of jungle which runs along the creek right through the middle of the farm. Of course, the locals have never seen any of the nocturnal animals as they do not have decent torches.

It was lovely for Pip to get emails tonight and she says thanks for so many birthday wishes. Just on dark was the first internet connection we had all week, as apparently the government turns off the internet in the whole country during the year 10 major exams. It is supposed to stop cheating but we have not actually spoken to any local who has used the internet let alone worked out how to cheat on it. We suspect the Government officials have probably no idea how it works either.

EE.17: More Frustrations – work ethic:

Early on, it was clear to Pip and me that how we managed our frustrations which inevitably arose would largely determine the success of our assignment. When faced with the different work ethic and business drive to which we were accustomed, we could have lost heart and retired from the front line gracefully. Or we could work on different techniques to achieve our ultimate goal of a productive farm within the context of maintaining the community at the farm and not forcing anyone into further hardship by laying off staff that wanted to continue to be employed at the farm.

We certainly allowed ourselves to "boil over" occasionally, so long as the foundations were solid and we treated everyone one with respect. At the farm, many management changes were required, but working out the best way, psychologically, to achieve those changes was THE most serious challenge. Should there be written submissions, or white boards, or meetings, demonstrations, orders, long term plans, or short term plans, monetary incentives, or personal incentives? Maybe all these things and more but how to know? Occasionally, we had to let our instincts take over and simply demand an outcome...

Day 122

I started the day with a try for internet access but still no success. We had many issues off the farm that we needed to resolve and it is incredibly frustrating. It could not have been a worse week to have no internet.

When I came back frustrated at no internet, I found a gaggle of managers wandering around shuffling paper, but mostly staring at navels. Probably not a good mix with my mood, I must admit. When a couple of heads appeared above the parapet, the tone of my voice was anything but cool, calm and collected.

Suddenly I was sending off managers in all directions cranking up some action. I asked when the cabbage seedlings were to be planted out (cabbages should be one of the major income earners for the year) and after a bit of scratching around Tibebu and Lechisa decided that it should be in five more days - as that will be thirty days from planting the seeds into the nursery beds - and the rule of thumb is thirty days.

"FIVE DAYS – how the hell are you going to have them in the ground in five days. Nothing is being done today, the next two days are the weekend, you have already scheduled management meetings for Monday and Tuesday, why is there compost not being spread right now? How are you going to prepare the paddock in time …… .????????? and more……." Strangely the population in the courtyard reduced by eighty percent, two men were sent off with shovels to load compost, Merka decided he also needed to get out quick and head for Ambo, the dairy manager, Getun, was summoned to drive the tractor and trailer (with my help to back the trailer) and, showing a lot of good grace, Tibebu came to me for a discussion and to hear what I thought should be happening. As usual we were on the same wavelength.

Then he told me that there was to be meetings on Tuesday and Wednesday with the view to reorganizing all the staff and their roles and responsibilities. This is exciting news for Pip and me. I asked if we could meet in his office to go through the current roles of all the managers, cashier, purchasing officer, store keeper, driver, foreman and others and give him the ideas that Pip and I had been formulating.

We have just discovered that there is a foreman under the overseer in the management structure, and he is meant to keep the labourers working. He is an unpleasant guy who looks shifty, and we have never seen him lift a finger. Tibebu as farm manager dislikes him, but Lechisa the overseer thinks he is useful, so that will be played out this week.

Chapter 6

June 2014 - Bureaucracy

EE.18: Driver's License

Ethiopian roads can only be described as chaotic at the best of times. Combined with highly unreliable and poorly maintained vehicles and many very inexperienced drivers, being on the road is a dangerous place to be. Every trip longer than an hour will almost certainly mean coming across a traffic accident and Pip and I witnessed the aftermath of at least two serious crashes on every trip that we took into Addis Ababa.

Our Peace Corps friends invariably travel to the city in the local minibuses, but we were very concerned at the likelihood of being in an accident if we used that form of transport.

By acquiring our own driver's license, we at least had our destiny in their own hands and were not relying on high risk taking drivers who had been chewing the hallucinogenic leaves of the Khat plant – so common in Ethiopia. In due course, we also acquired our own vehicle – an old Toyota Land Cruiser. We purchased this vehicle from an Australian couple, Patrick and Katarina, who had been living in Ethiopia and were returning to Australia. By coincidence, Patrick is distantly related to me though cousins.

Given the poor state of the roads, lack of reliable maintenance and the general harsh operating conditions for vehicles in Ethiopia, the Toyota Land Cruisers and Hilux Utes are definitely the favoured vehicle. The staff from the numerous Embassies in the country and the large NGO population invariable travel in Land Cruisers.

Day 126

Sometimes the luck is with you and today was one of those days.

Yesterday we travelled into Addis so that we could begin the process of getting our drivers licenses. After breakfast at the Kaleb Hotel, we took a taxi to the Hilton Hotel where the Australian Embassy is based. The first stage of our license application was to get a letter from the Embassy saying that our current Australian license is valid. We found our way to the right lady in the Embassy and paid our three hundred birr ($15) each for her to write the necessary letter. Luckily (as it turned out) she did not know the location of the second stage of the application which is the Ministry of Foreign Affairs office. The Embassy Consul, Stuart (whom we had met at the farm a month ago) was passing the office and came to us for a chat. He also did not know where the office was, so we headed off with him to find his local driver.

We passed by the Ambassadors office, and she called us in for a lovely informal chat for a few minutes. It was much easier than when we had met her months ago for our formal introduction.

We then found the driver, but he had trouble explaining to us where we should head for stage two, so Stuart told him to take us there in an embassy vehicle.

Like most Government offices we have seen so far, the immigration offices building resembled a rambling tin and concrete shed similar to what you would find for the club rooms of a very poor country footy club. Lots of service windows – mostly at the top of concrete steps and some with queues of people. Even the driver had trouble finding at which window we should present, and we went back and forward, up, and down for quite a while until we handed over our letters and another three hundred birr each through some steel bars. And then we waited.

After lots of signing and stamping by two people in the office, we were given our receipts (all in Amharic of course so no idea what they were for.) It is unlikely we would have ever found the right window on our own - but I guess something always turns up.

Back into the car and out onto the road and the driver told us he did not have time to take us to the Ministry of Transport office for the final stage of our license, as it is right on the other side of the city. He hailed a taxi for us and negotiated a price of two hundred birr (A$10) which was very cheap as we had expected it to be four hundred. We drove for well over half an hour right around the city. You must understand there is a ring road but as there are sections missing, you push through crowded streets until it starts again. Also, the drivers often try and take short cuts which I think take much more time but maybe are slightly shorter!!

Eventually we arrived at what looked like an old wartime airport with huge steel hangar type buildings and old dilapidated concrete offices amongst trees - mostly Australian natives.

We paid the taxi driver, and he said he would park and wait for half an hour for us to take us back to the hotel and if we were longer, he would head off. There were masses of local people about and before we had even entered the building, we were beckoned by a guy who wanted to see our papers and then thought we should pay him some money. Pip was quick enough to realize that he was a fraud, but from then on it made us suspicious of everyone.

With a few questions directed to people who we thought may speak English, we found our way into a huge tin shed with thousands of people milling about, masses of windows with queues and not a sign to be seen. A guy came up to us and directed us to a small man who was using an upturned 44-gallon drum as a table. He checked all our letters and papers and told us to go to a window to get some stamps – similar to postage stamps - to show that you have paid someone some money.

Then back to the drum where he filled in our forms for us – all in Amharic, and then another fellow appeared and started to assist. We figured all of this must be legitimate being right inside the building but with the area thick with humanity we could not be sure. This new fellow then took us around to another window where only a few people were waiting. It turned out that our new "assistant" was our taxi driver – and we didn't realize this until much later.

He must have decided that we looked pretty clueless and with minimal Amharic would certainly need help, so he had appeared at our elbow. Anyway, the office ladies finally got to our forms and were filling them in when the clock hit 12.00noon (or six Ethiopian time as the clock had started for today six hours ago – at 6am). Time to knock off for lunch for an hour. So out we went to try and find some lunch. The taxi driver said he knew a place over the road- so we held our breath crossing the very busy road (the main road south of Addis) and found a cafeteria style buffet lunchroom.

An hour later we were back in the Ministry of Transport office in the old tin building, still swarming with people but now we were somehow at the head of the queue. After another half an hour we had two drivers' licenses- but alas - they both had my picture laminated on them!! So back again for another try and in ten minutes we had one license each. Bingo! We can now drive in Ethiopia. Thing is, we are pretty reluctant, as the potential for an accident is very high, and if you hit someone as a foreigner you are always liable regardless of who is at fault, and you can get beaten up by the locals in the area if you are lucky - or go to prison if luck is not with you. By mid-afternoon we had achieved what seemed impossible and the locals said it was unheard of to have a license in less than two full days!!!!

We took the same taxi back to the Kaleb Hotel where we paid off the taxi driver with a large tip to thank him for his assistance almost all day. We were in time for an Australian volunteers get together dinner at the restaurant next to our hotel. About twelve of us squashed around a table and it was very noisy - but great to catch up and meet a few new people. A very tiring day so early to bed.

EE.19: Bureaucracy

So many aspects of the Ethiopian economy and culture conspire to restrict or at least slow development and growth of the economy which is so desired by the Ethiopian Government and its people. Whilst corruption is rumoured to be ripe in the upper sections of society and politics, there are numerous efforts to prevent fraud and corruption throughout the lower levels of the economy. Unfortunately, the complexity of red tape which results, very often means misdemeanours are harder to see and loopholes are left open everywhere for those who know how to step through.

At the Guder farm, most of the financial systems that were in place had been dictated by the heavies at head office and often they were not appropriate for the farm situation. As Pip and I gained a greater understanding of the systems over the first few months, it became very obvious that the financial system and record keeping – though complex and demanding – could easily be manipulated by managers to suit their own desired outcomes. Meanwhile the whole system was weighed down by the overly bureaucratic structure.

So often in farming timing is critical. Very often activities at the farm could not be commenced or completed because, to purchase the inputs, quotes had to be gathered and then money had to be "requisitioned" from head office in Addis with personally signed forms. The purchasing officer would then be sent by the minibus system to Addis to personally hand in the necessary forms and he would then return to the farm a day or two later.

In due course the money would show up in the farm bank account in Ambo and a special trip would be made to the town with the cashier, who was the only person authorised to withdraw money from the bank. The chance to respond to opportunities for good productive activities had been lost due to poor process and the length of time required to progress things.

Day 139

Our first trip to Addis by public transport.

We left the farm early and walked the seven kilometres to Guder with our backpacks on, accompanied by Tibebu. Then in a minibus from Guder to Ambo. This was the worst part of the trip. There were twenty-one people in the twelve-seater, and we prayed that it was not going to be like this all the way to Addis Ababa.

In Ambo, Tibebu chose a minibus for us, and our packs went on the roof. There are three levels of buses for most routes, and you can choose which you want and then wait until it is full so it can set off. Tibebu likes to choose level 2. He says that Level 1 – the newer ones go too fast and have lots of crashes. Level 3 either break down or crash because they do not have brakes. The price is the same for each category, so it is all a bit of a mystery to us!!

We departed in our level 2 soon after we hopped on but had to wait for ages to get out of the bus station in Ambo as the entrance is only 1 bus wide and many buses came in from around the blind corner, so we had to back out of their way a number of times. Good idea, Eh! A bus station with buses coming and going a dozen times a minute, with only single lane in!!!

The driver picked up and dropped off quite a few passengers on the way to Addis Ababa and finally arrived in the chaos of the Merkato bus station. Immediately we alighted we were swarmed with street sellers and taxi drivers.

It was still quite early in the day, so we took a taxi to AGOHELMA head office, which was quite close. There we were greeted by Gayle (new AVI volunteer from Australia) and had a good chat with Elias (local public relations employee of the organization with good English) whilst we had lunch of injera at the cafeteria. Now that Abebech Gobena, the founder of the organization, is in failing health, the organisation is run by two general managers, Ato (Mr) Eshetu and Ato Tesfaye. We met with Ato Tesfaye about the purchase of a vehicle and also about purchasing the equipment and supplies for the chicken caravan that I planned to build if we could obtain an Australian Government grant.

We found out that for purchases to comply with the organization rules, we needed to approach three businesses for each item and get three sealed and stamped quotes (proformas) for items over five thousand birr (A$250). These are then opened in front of a committee from the finance department and the best one chosen. It is obviously not going to be easy to buy these things - and buying them with our own money and claiming the money back is not possible either so we are told. The paperwork must be present for the organization to provide evidence to the auditor.

Finally, we got an AGOHELMA organization car to take us to the Kaleb Hotel for the night. We had dinner at Edna Mall - just up the street from our hotel, and then back to our room to watch BBC and CNN news on TV to catch up with the world news - such a luxury. English Premier League football is big in Ethiopia and nearly everyone has a team that they follow passionately on the few TV's that are around.

Day 150

For the purchasing exercise today, we were given a vehicle and driver from the organization, and they picked us up after breakfast and then also picked up Tibebu who had stayed the night in Addis. We headed to the part of the city where all the tool/hardware shops are located in Piazza area, near the old Post Office building. We selected a welder and Bosh angle grinder and proceeded to get 3 proformas (quotes) in signed and sealed envelopes from 3 different shops.

We priced some other smaller items and would have happily purchased them there and then with our own money, but such was not to be. Back at the AGOHELMA head office we attended the meeting in the finance department office where, after each of the four "important" people had been assembled from their different offices, the unopened envelopes were carefully signed. They were then ceremoniously opened, and the quotes revealed.

Lots of note taking and signatures were required on every page of every quote with important looking rubber stamps being thumped onto everything flat enough to receive the numerous ink tattoos - but alas the winning quote could not be revealed to us.

A purchase order was then filled in with more multiple signatures and rubber stamps, and the order document was then taken to the person who orders the money from the bank. At this stage we could see no end to the paperwork, and it was getting late, so we left them and got a lift back to the main part of the city centre with a group of American school students and their teachers, who were travelling in Ethiopia undertaking charity work. They had just competed scrubbing and painting the walls of the orphanage kitchen at head office.

We were dropped at their hotel, and then we walked a few kilometres back to the house of our Australian friends Patrick and Katarina. Patrick has a contract with the Ethiopia Agricultural Transformation Ministry and Katarina is with the International Committee of the Red Cross. They have become very good friends and have taught us so much about Ethiopia. They are the people who inspired us to get our drivers licenses and to explore the more remote areas of Ethiopia.

EE.20: Moral Dilemmas

Living so deep inside a completely different culture was always going to present us with moral dilemmas which would have to be faced. So often during or assignment we would just have to accept what was going on, react as best we could in the circumstances and continue to show appropriate care and compassion.

Day 159

Back at the farm and this morning we were out of bed in good time as we had been told there would be sweet corn to prepare for a trip Ambo. By 8.00am there was a large mound of cobs that had been carried up from the fields and dumped outside the ute garage. Apparently, the guards had been told to pick cobs overnight and leave them in bags ready for the labourers to carry up to the farmyard this morning.

Dairy manager (and trainee driver!!) Getun packing corn ready for market.

The cobs were then tidied up with the end broken off and outside leaves removed just leaving a few leaves around each cob and any cobs with bird damage removed and put aside for sale to local community women. The quality cobs are then stuffed into large bags that carry about seventy-five kilograms (about 220 cobs). These are covered in a bit of green grass to keep them fresh, the tops of the bags sewn with nylon thread and the bags loaded into the ute.

Tibebu was keen to get the six bags to the merchant in Ambo as soon as possible so we headed off with Pip driving and Getun in the front passenger seat so that he could start to get an idea of the driving techniques from Pip. He has a driver's license but has never driven a vehicle on the road - so quite a challenge coming up.

We dropped off the corn in a side street at the same place as yesterday and again received the good price of two hundred and forty birr (A$12) per bag.

Once we got past the worst of the crowded road in Guder on the return journey, Pip stopped and Getun took over as driver. He had a reasonable grasp of the clutch and brake and road position amongst the other vehicles, pedestrians, and animals. Unfortunately, sometimes he forgets to turn the steering wheel so a few yells are required at times!!

When we got to the gate into the farm, he panicked and hit the accelerator and the brake at the same time. After some anxious milliseconds, we stopped inches from the gate post on a strange angle to the gateway. Apart from a bit of pride, luckily no damage was done.

Son Alistair and his partner Sophie have come to Ethiopia for a visit and are staying with us at the farm for a few days. As we are having a few problems with our computer Ali decided to reinstall Windows but as so often happens with his efforts to "improve" our computer, there was a complete failure of everything. The virus protector had deleted the boot program. I doubt there is anyone within a thousand kilometres who could have solved that problem but, by tonight, sure enough Ali worked his magic again and all is good.

Meanwhile Sophie was sitting having her long hair braided by Zinash, certainly a very important activity for the women in Ethiopia culture.

Today there were fifteen day-labourers working on the farm, helped at times by a few managers, so we were delighted to have a hive of activity. The ladies that helped with the corn then went hoeing cabbages which are just starting to get lovely looking hearts. Four men were driving the four pairs of oxen preparing ground for teff grain to be sown and the rest were sowing teff in a field already prepared.

I should have said twenty are working today as of course the three cow boys were hanging around the cow shed doing their usual couple of hours for the day (paid for 8 hours) cleaning out the shed and there were two shepherds looking after the cows.

Just before dark one of the lovely ladies who is a day labourer appeared near our hut in floods of tears and she collapsed to the ground hysterical, clutching at different parts of her body. She has no English, and I was the only one about and was not sure what level of comfort to provide. Pip came quickly and called Zinash, and a few managers and workers appeared soon after. We were left to guess what was going on. My reaction was that she may have had a miscarriage, and Pip was backing that she had been raped by an unpleasant member of the staff - such was her distraught demeanour.

After a short time, she was taken by other managers, and they all disappeared with her back out to the farm area where she had been working. It turned out that she keeps her savings with her in a pocket of her dress (so it does not get stolen from her home) and today while hoeing she had lost the four hundred birr (A$20) that was to buy fertiliser for her crops to feed her family for the next year or so. One can understand her devastation!!!.

We are not sure what the outcome will be and hopefully it will be found tomorrow. Pip and I would love to just give her the money but that creates serious moral dilemmas in the world in which we live.

If we give a cash handout to one person, what happens when the next disaster occurs? It is not good for us or our relationship with the community if we are just viewed as cash cows. After discussing it with Tibebu, we gave him the money, and he said he would give it to her surreptitiously from "the farm".

Another day of having water and electricity all day – quite miraculous as we usually lose one or the other each day - but greatly appreciated whilst we have visitors. We were all invited for dinner with Lechisa and Zinash which included lots of injera – in fact crushed up injera cooked with onion. The dish is called firfir, and it is put with a boiled egg on top of normal injera. We all tackled it bravely and enjoyed the cheesecake that Pip and Sophie had made earlier in the day.

Day 165

I headed off to Guder early to pick up Tibebu and the men from the electricity company who had said they would come and perhaps be able to increase the voltage from the local transformer. This may mean that we can get the irrigation bore going which we have not yet achieved – but is a very high priority.

Previously when I had managed to get an electrician to the farm, they found that the power at the switchboard for the bore was only 350 volts (instead of the "normal" 415 for 3 phase power). I had emailed this information to an electrician friend in Australia who is helping us remotely with our efforts to get the bore operating. He said that based on all the previous information I had sent him, we would need at least 380 volts for the bore to work.

He had also convinced me to purchase a voltmeter and with that information, he remotely guided me through testing this and that, and he reckons that the pump down the bore is OK if we can get sufficient power to it.

We dropped off one man at a power pole where the power to the whole district could be switched off, about halfway between Guder and the farm. When we got to the farm, the other guy phoned his mate and got him to disconnect the power. He then climbed up on top of the transformer and twiddled a something. The result was that we now have 385 volts and are praying that it may be an easy fix for the bore.

After ringing his mate who was sitting up the pole half way to Guder and getting him to turn the power back on (no doubt to the relief of every power user for the fifty kilometres up to the village of Shenen in the mountains who had been shut off whilst we got twiddled), I brought the same guy from the transformer up to the switchboard. He found that two of the three circuit breakers where the power comes in were not functioning, so we will need to get a private electrician to come and replace them and connect the wires back up to the bore. Then for the big test!!!

Getting water out of the bore is definitely the number one priority of head office management for their Australian volunteers!!

We loaded up with another six hundred kilograms of corn cobs into the ute and Pip headed for Guder to drop off the electricians and then on to Ambo with Tibebu and Getun to sell the corn. When she returned, we loaded up again and I took in the second load for the day again with Tibebu who does the negotiations with the merchants, and Getun who helps fight off the thieves who try and pinch the corn out of the bags when we stop in town at the incredibly crowded market.

Pip about to drive the ute loaded with corn down to the market in Ambo – with Lechisa on hand to supervise the loading. The local village women have come to buy the rejected corn which has been damaged by birds and cannot be sold at the market.

Getun also helps unload and we let him drive home after we are past the worst of the road and traffic. There is a long way to go but hopefully he will be able to drive to Ambo one day.

Later in the day we reloaded with corn, and I did the third trip to Ambo. On the way back we stopped in the main street in Guder whilst Getun went off to buy something. A man dressed in rags came up to my window (there is nearly always someone who comes up - if not many) and he started begging. He then pulled up his trouser leg and revealed his leg which had a huge dirty gaping wound, and the bone was visible for at least 3 inches.

I have become pretty accustomed to seeing human hardship, but this really took my breath away. As usual that dreadful moral dilemma raised its ugly head. I could give him a few hundred birr

(A$30) which would mean nothing to us and probably mean that he could get it properly treated and it would heal well. But if I did that, what of the need of at least another thousand people within a few hundred meters with their own varying levels of hardship – potentially many as bad as his. Any sign of me handing over money would have meant the ute surrounded with people begging and all future visits to town would be a nightmare, as Pip and I are so well recognized being the only whites within a hundred kilometres. But could I just turn away from a fellow human-being in such distress?

I muttered “not today” wound up the window and looked the other way – but only from this poor individual and not from needs of this country which are becoming etched into my soul. It is very challenging on so many levels, but we are becoming accustomed to such dilemmas now. The first one thousand beggars you knock back in Addis Ababa – particularly the deformed, the old, and the young ladies in rags with babies are the hardest!!! Is it even possible to think like that?

On a brighter note, we had a good laugh today. Lechisa needed another injection for his malaria. He had been sent home from the local health clinic in Guder yesterday with a syringe full of medication and a cross marked on his backside in biro where the injection was to go. It was suggested that I might do the deed but there is no way I will get involved in injecting anyone other than Pip if necessary.

Zinash was quite happy to administer the injection, but Lechisa is so black we had trouble finding the cross! We had to bare his backside outside in the sun so we could all search for the cross. Eventually the injection was administered and Zinash was delighted with her effort. She loved me calling her Dr Zinash!

EE.21: Traffic in Addis – and Reviewing Farm Finances

As the year progressed, the farm driver Merka took a job in Ambo so Pip and I resolved that we would act as drivers until another driver was found – or we could train up Getun who had a license but was still just learning to drive.

Having our licenses meant that we were starting to drive to Addis Ababa every few weeks, learning how to find our way around the city without maps and the learning the etiquette (rather than road rules) of driving in this country.

Day 172

We were up in good time and packed ready for the expected departure to Addis Ababa in the ute at 8.00am. Years ago there were consultants in Australia that were called "time and motion" study experts. I am sure they could do a full PhD centred on our progress between 8 and 9.00am. We are so accustomed to it now; we just wait it out. People come and go, people get in and get out of the ute, things get loaded in the back, managers appear looking to get things signed and everyone heads into an office. Finally, in our courtyard area all is in order, people are all inside the ute - and produce in the back.

Pip starts up the ute, but then there are orders from someone to go and pick up more produce that needs to go to the head office in Addis Ababa from the bottom sheds on the farm. We drive the one hundred meters down to the other shed, all climb out, and the loading up begins again – we had no idea that butter, cheese, teff and more had to get to Addis on this trip.

The rope and tarpaulin are found as the weather looks inclement, but of course not put on at that point despite the threat of imminent rain. It is so much better to do it in the rain amongst the traffic somewhere on route!! Everything could have been loaded an hour ago!! Finally Pip and I and Lechisa are joined by a quartet or more of others who want a ride to Guder or Ambo and the trip to Addis Ababa begins.

Tibebu joined us in Ambo in the pouring rain, and we went around to his house in the university campus and dropped off some teff flour. After that we had a pretty uneventful trip to the city. I drove the last bit into Addis and to the AGOHELMA head office. Constant rain for the whole trip did not make it any easier. The thought of the canvas showing through on the tyres, which are way past the bald category, gives me some trepidation on the wet roads!!!

Addis Ababa traffic is quite indescribable and though it seems impossible there are actually no road rules that we have been taught or have worked out for ourselves – except that, where possible, you should pass oncoming traffic on the right (opposite to Aust). Apart from that – nothing. All intersections are completely uncontrolled, so it is just a matter of bluff. Yes, even big intersections with 4 lanes or more coming from a number of different directions are just a free for all. You would swear it couldn't , but it does.

There are large roundabouts in some places and mostly cars go around the right way (anticlockwise) but sometimes if it looks a bit busy that way, you duck around the opposite way and find the exit you need!!!

Roadworks often mean that one half of a divided road will be blocked off with rocks but there will be no signage whatsoever, so traffic suddenly starts going both ways on the other side of the road. Animals and people constantly wander all over the road just like we experience in Guder and Ambo.

The line taxis (minibuses) and old Lauder blue taxis duck in and out of the traffic with only inches to spare, always on the lookout for a potential client on the side of the road- so their stops are very unpredictable.

Of course, the roads are almost always broken up with huge potholes. If anyone wants to put a pipe under the road from a relative's house to their house for example, they get a pick and shovel and dig up the bitumen, dig a trench, put in a pipe, and fill the trench with dirt. Inevitably in no time there is a huge trench in the road that has all traffic almost stopping to negotiate it – with consequences for traffic flow for miles around.

Actually, there are a few good quality roads now, including a ring road which is almost complete right around the city. Many of the worst areas for traffic flow and road conditions are a result of the new suburban railway lines being put in from north to south and from east to west. As it is all being constructed at once, the effect on road traffic is diabolical. Being the wet season there is mud everywhere and the big trucks leave huge clods of dirt on the bitumen roads to add to the slipperiness.

We arrived safely at the head office at about 12.00 and had injera for lunch in the cafeteria. Then we were ushered into a manager's office ready for the meeting with the head office managers and finance people. The aim was to seek to improve the financial performance of the farm, and particularly the dairy enterprise.

Tibebu, Lechisa, Pip and I representing the farm, settled in with a couple of the head office managers and started to wait for another half a dozen finance people and Abebech Gobena herself to arrive. Given there was only one spare seat and at least six people still to arrive, Pip and I were both wondering – wondering – wondering.

Sure enough, when another half dozen people arrived it was decided that we could not fit in that office, so we all moved to a small meeting room. (In that half hour wasted, the Chinese had no doubt built a billion widgets and the Wall Street had turned over a trillion dollars of companies bought and sold).

The meeting went for three hours and was all in Amharic but Tibebu or one of the other managers would interpret for us at important times. The point of major consternation (which we had no idea about previously) seemed to be that three years ago on Tibebu's insistence, head office had given the farm five hundred thousand birr (A$25,000) as a capital injection to try and make it more productive. It seems the money has all been used up with no increase in production and the question to Pip and I was "can it be made profitable".

Fortunately, we had spent quite a lot of time on previous days preparing a large document with all the problems we see in the dairy enterprise, and we offered a solution for them all. I was also at pains to explain that it would be at least two years before they would see major change in the productivity of the dairy enterprise as that is the nature of farming, but we expressed confidence in a good result.

I could also assure them that in the short term they would see greatly increased returns from the crops - so I sure hope the season continues to do the right thing by us. I think they will be smiling next time we meet.

Our feeling is that the money all got used up in the "retirement village" atmosphere the farm had when we arrived. Mostly it went paying what they described as wages, but we would describe as pensions!!! It will be interesting to see if it gets more mention or if everyone has moved on and Tibebu has managed to explain it away with some creative accounting. In his defence he did argue that the value of the dairy herd had increased during the last two years, which is true.

I had a private meeting with the head of finance (who seemed pretty switched on) and got a much better understanding of how they keep the accounts, and he has given me a proper profit and loss statement for the farm according to his books which is very, very different from the rubbish figures that have been presented by Tibebu at the organization biannual meetings.

EE.22: Local Involvement:

Wherever possible we said yes when invited to community events and the local people were always so welcoming. We did have to be very careful not to drink the local water as inevitably it would lead to gastric problems and so the first thought when we headed off to local celebrations was to take our own water bottles full of water that we had boiled.

Day 177

What we had expected to be a quiet Sunday back at the farm turned out to be anything but. I had planned to do some bookwork and get my Australian BAS lodged to the ATO for the last financial year. First, we decided to go for a walk to see the beautiful set of waterfalls on the river which is the back boundary of the farm. We expected them to look spectacular as we are now in the heart of the rainy season.

We passed by Mr Ollie, our neighbour's neat little farm with three tukuls for the home of his large family. We get on fine now with his children who had thrown rocks at us a couple of months ago. With the relationship repaired, there is lots of waving and hellos and goodbyes as we pass.

After all the rain, the waterfalls are impressive, and the silt laden brown water causes brown spray which discolours the jungle quite a long way up the valley walls. We were lucky enough to see a troupe of eight beautiful black and white Colobus monkeys that live in that area. We walked through the jungle/forest along the top bank of the valley. This area is part of our farm and supports about twelve hectares of our eucalypt plantation, but it looks like jungle as it has a heavy understorey of thorny shrubs and vines stretching up into the trees.

One of the troupe of Colobus monkeys down near the river.

As expected, we came across various groups of kids, picking up any sticks and branches they can find to take back to their family for cooking. They make little stacks and then carry them on their backs along the winding paths that have been created over the years.

We returned from our walk and were just thinking about lunch when Lechisa told us we were invited to a kid's birthday party which was about to start up the road. We had no idea what to expect and we had other things to do, but we now have sharp instincts, and we answered yes instantly. Fortunately, we got a bit dressed up and gathered up a couple of bottles of soft drink.

Up the road we walked with Lechisa and Zinash and into a hut beside the road. The yard was the neatest we had been into with flower gardens, beautiful vegie patches and obviously the farm behind very well managed. There was a new house not quite finished with a corrugated iron roof and superbly mudded walls and even some glass in a couple of windows.

Inside was the Mum and Dad, lots of their children and the grandfather. We knew the grandfather as he is one of the large important landholders in the area that Pip and I call Kings. Obviously, his industriousness had rubbed off on his son as this was the best-kept property in the area. The King has no English but is a nice chap and is reputed to have/had about twelve wives.

We had not met the Dad before, but he looked a nice fellow and obviously was a very hard worker. With just the family members and Lechisa, Zinash, Pip and I the celebrations began.

Green branches had been brought in and made to look like a Christmas tree which was decorated with a birthday sign (in Amharic), balloons and a few biscuits and lollies hanging on it. Large bowls were filled with popcorn and roasted barley and the table had fruit, soft drinks and large flat round loaves of the local bread which is made from the flour from the false banana trees. It turns out it was a birthday for the four- and six-year-old boys.

Not many people really know their actual birthday in this area, but this resourceful Mum has decided it works well to have a party now and again for the boys and call it their birthday. Their five older sisters and the mothers home helper (usually a girl about twelve years old "a relative from the highlands") busied themselves handing around the food and making coffee and, with Zinash's help, setting up candles and photo opportunities.

Pip shot home and got our camera and a local boy had been invited to come and take photos as he must be known to have a camera. The boys and girls (all gorgeous looking kids) slowly got over their shyness caused by the two huge white "farangi" in their midst and ended up sitting on our knees and playing with my funny big wrinkly, white hands which always seem to cause amusement with the children.

At a local birthday party

At one stage the Dad and Grandpa got a message and headed off. Lechisa told us that one of their guards had apparently been attacked and had a head wound. We did not see them again and not really sure what the story was, but this is Ethiopia so you never quite know what might have happened.

After a couple of hours eating and drinking, we had a guided tour of the garden and walked hand in hand with some of the children back down the road to our farm.

Truly an afternoon to remember. We keep thinking how lucky we are as you could never "buy" such an experience – even from the most expensive tour company!!!

Late in the afternoon Pip went to the paddock to get the internet. She took her pet Mrs Sheep with her for an outing thinking she might like to eat some of the feed in that area of the farm, but of course she was only interested in eating the dongle and licking the computer screen and generally rubbing up against Pip - like the dog that she pretends that she is!!

I did our weekend washing which will take a couple of days to dry I expect. Put the stew on for dinner early hoping that a couple of hours of boiling the meat may make it chewable but the meat we get from the butchers around here is not going to be defeated so easily unfortunately. Tomorrow we may go to the market and buy a boyfriend for Mrs Sheep and another sheep to kill for our rations. It is so much better meat than the beef we can buy at the outdoor butcher in Guder or Ambo.

Chapter 7

August 2014 - Rural Ethiopia

EE.23: Bore Success – Celebration:

Gradually we started to chalk up some successes and whilst the dairy output was beginning to increase, and the crops were looking promising, it was water flowing from the bore that caused the greatest excitement – both at the farm and back in Addis Ababa at the Head Office. I imagine many years ago, when funding had been achieved for the bore project, and the bore had been fitted out, there was great excitement. Then when the button was pushed, and no water appeared, all the worst fears of mechanization in Ethiopia were realised. Another expensive "White Elephant" sitting there year after year.

Day 185

We are sitting here with a warm inner glow as I type up today's activities, and we are quietly dropping off to sleep in our chairs.

I was up early to go to the cow shed and check on whether the latest round of management changes is operating correctly. Yes, the cow that calved a few days ago had her calf getting a decent drink and then she was milked out. Finally, we may have got that one settled.

Having thought yesterday we would go to Guder first thing to buy two sheep, we were told that we should wait here for Tibebu before setting off. We thought that could mean anything, but he appeared on the local bus from Ambo soon after and was ready to head off down the road again almost immediately. We set off with Tibebu and Lechisa – both apparently to help us buy our sheep. We thought that was to be a quiet trip but, by the time we had driven to the bottom farm gate we had eleven people in and on the ute!!!

Pip drove us all into Guder and we bought two sheep in the market - A husband for Mrs Sheep (Mr Ram), and another lovely ewe with a bit of age for us to eat. Mr Ram is light brown with neat hair, a deep rammy 'baa' and very gentle and easy to lead. We needed a husband for Mrs Sheep as despite the early confidence from all and sundry that she was pregnant when we purchased her, that turned out not to be.

The market was even busier than when we were last there and it was just a matter of pushing our way through the hordes of people, sheep, and goats, finding animals that suited our requirements and then getting Lechisa to see what price he could buy it for. We already had our ropes in hand to lead them away and tie them into the back of the ute.

Having made our purchases and paying our seven-birr (30 cents) tax to a guy at the gate we headed for the ute. Pip had a quick visit to the donkey market and was surprised that prices were not much higher than for a sheep. Age, not size, seems to be an important factor. They never kill donkeys- they just grow old and die.

On the way home we called into the local welder and found out his son was an electrician, so we organised to pick him up later in the day to come and see if we could get the bore running. After we got back, I killed and hung up the sheep and then went out to the internet to download Grade 10 national exam results for lots of the local children who had no other means of receiving their results.

Pip got a call from Tibebu- come now to pick up the electrician. So, she headed off again for Ambo in the long-suffering Hilux ute, picking up Tibebu in town and then on to collect the electrician. Pip queried his lack of tools, so they detoured to his house to pick up a screwdriver and pliers!!

When they arrived back at the farm, the electrician put his tester on a few points and declared that a large switch was not working so he may as well go home. I was determined to get somewhere so suggested he bypass the switch with some pieces of wire which I supplied so we could at least test and see if the recently negotiated increased power supply to the farm would do the trick.

Water – precious water - from the bore and a celebration by senior farm staff members.

Finally, he connected up the irrigation bore pump and BINGO - we got water!

There was great elation from all the staff, cheering, yelling, and dancing - especially Lechisa and Tibebu who immediately phoned Head office and reported that they responded by "crying"! Then the electrician connected the welder and "bingo" that worked (although I later reorganised it and wired in an earth as the electrician said it did not need an earth on it despite what was clearly written in the instruction manual).

Pip then drove the electrician back to Ambo for her third trip to town for the day - picking up a policewoman who needed a lift to Ambo. Three trips to town on a Monday market day negotiating the human/animal tidal wave without an incident. My girl is good.

This evening, we got a phone call from Tibebu in Ambo to say he had found some limestone which can be regularly mixed in with the cattle feed. For months we have been adamant about the need for this in the diet to increase productivity - so that is another dragon slain today.

With so much progress for the day, Pip and I went in to the hut of Lechisa and Zinash's with some Fanta and Pringles to celebrate. They told us it was a local children's celebration day and, in the evening, all the community children light fires, crack whips and join together to visit houses and sing for money.

We headed out onto the road with Lechisa and went to a couple of fires in house yards of the village and chatted to the occupants and kids who were hanging around the fires.

Then a group of young local boys came into the yard of one hut that we were visiting and sang for us - one leader chanting something, then the others chanting and singing in response. It would have been lovely to get a recording as it was incredible and seemed so well planned and practiced. Apparently initially they chant about – "by giving us some money, luck will be with you for the year". Then after you give a few birr, they sing/chant about how "God will look after you for the year and your cattle will get fat." We gave them some birr (in the hope that God may also look after Mrs sheep and her new husband), then headed back to cook some dinner. So, what a day! Many dragons slain and a wonderful Ethiopian cultural experience to top it off.

EE.24: The Ethiopian Calendar and Clock

I often wondered what went through the mind of a Canberra bureaucrat processing our fuel receipts. When AVI asked us to drive into Addis Ababa for official business, we'd be reimbursed — but the receipts we submitted told a different kind of story. I'd glance at the date and chuckle: 3am, on the 4th day of the

13th month, 2007. In reality, we'd filled up at 9am on the 4th of September 2014. I always pictured someone back in an air-conditioned office in Canberra, eyebrows raised, wondering whether to question it — or just quietly move on.

Day 192

Out to the internet early as I was keen to get a photo to head office of us all around celebrating the water coming out of the irrigation bore. Tibebu had taken the bus to head office in Addis today, and I thought he would love to be able to show all the managers the triumphant scenes.

Pip went off to plant some native Podocarpus trees and was lucky that she caught up with Lechisa along the way and spent an hour or so with him getting him excited about planting more native trees about the farm and having him dig the holes for her. Along the way she had a good chat to him about the Ethiopia Calendar as it is nearly the end of this year.

We have 14 days until the end of this month of August, then there is a five-day, thirteenth month, called Pagume – because every other month of the year has exactly 30 days. Then it will be New Year, and we will start 2007 on the first of September. Why 2007? Well somewhere along the line Ethiopians believe that Christ was born seven years later than what the rest of the Christian world believes – so 2014 in our calendar, is actually 2007 in Ethiopia.

Now just in case that has not done your head in, you must understand the way the clocks work. The clock starts twice each day – at 6am and at 6pm. So invariably when we are told to be at a meeting at "four morning", that means 10am, as the clock has started at 6am and the meeting is four hours later.

What time we get to celebrate New Year's Eve I am not sure but as always, we will just go with the flow. Incidentally, public servants do not get paid during the thirteenth month, so work output is less than ever for those days!!

EE.25: Limestone for the Cattle – to be or not to be

As time went on, we became more certain about what was required to increase the productivity at the farm and slowly we tried to tick off each of the things on our list. We became very familiar with the term "slaying the dragon" or permanently overcoming a problem– but this being Ethiopia would the dragon rise from the dead? Yes, quite often.......................

Day 195

The weather this morning was a bit miserable which is most unusual, so there was no chance of doing any welding on the chicken caravan which I have now started building, thanks to a grant from the Australian Government. I was keen to go to Ambo to pick up the bolts I needed for that project, and I needed to take the gas cylinder from our stove in to get a refill. Farm cashier Kidist is still in hospital, so Lechisa and Zinash and the farm cheese maker Tyto were keen to go and visit her.

We headed off after an early morning coffee - thank heavens I had the coffee and a biscuit.

In Ambo, I dropped all my passengers at the hospital, and I waited in the ute. I cannot bear to go inside as it so depressing and on almost every occasion we had heard of someone going into that hospital for treatment, they did not come out alive. We were desperately worried about Kidist being in there. Our dear friend – we could only hope for the best and we were heartened to hear she is improving.

Soon after Tibebu and Lechisa came out, and we went to do my shopping.

We picked up bolts, fuel, wire and mesh and the usual bananas.

As I pay for the items that I need for the chicken caravan from our own money, I will claim the money back from the organization having successfully negotiated this with head office despite initially being told it was not possible. The money has come from a grant that Pip and I were successful in obtaining from a special DFAT fund and we know it has been paid in full to AGOHELMA.

Reimbursement requires getting a receipt for whatever I buy (rightfully) but in this country that is a major undertaking. Normally it takes about 30 minutes, and I have no idea why - but it is always the case.

Occasionally it is done on a computer but mostly by hand, all in Amharic, but I can read the figures. After the receipt is issued then we all count the cash two or three or four or more times. None of the shops ever have the correct change so the next drama begins, and the shopkeeper or cashier just looks helpless. We all see who can look the most helpless which is quite a comical sight – are white people more capable of looking helpless than black people!!! Eventually someone goes to a next-door shop and finds the 2.5 birr change (15 cents) – which frankly I couldn't give a stuff about but seems to cause even more consternation if not paid or collected.

I went back to the hospital to pick up the others, and I distributed bananas to everyone in the ute (it is becoming a ritual). I pestered Tibebu again about the limestone and Pip had wisely sent me with a big wheat bag just in case.

Ah! he says we will try at the mine/factory (halfway between Ambo and Guder) as we go home. Up until recently he has been telling us that he has not been able to find limestone for the cattle anywhere in Ethiopia. Pip and I had frequently pointed out the quarry/mine/factory that we drive past every time we go to Ambo – but up until now to no avail!!!

We got to the gate at 1.30pm but the guard did not seem sure if we could get any product as it is lunchtime until 1.00pm. Seemed to my simple mind that maybe lunch would be over but apparently not. Tibebu finally headed off to an office after four staff had wandered over and handed our bag around to each other. The ladies I had in the back seat of the ute had all got out for a pee – just squatting on the grass near us, and I realised they had been sitting on the bags of bolts I had bought. Truly I do not think they feel discomfort or will just never complain. Such beautiful stoic women.

Tibebu finally reappeared having paid fifty-seven birr ($2.50) for half a cubic meter – the minimum they sell. Huge consternation, as the one bag we had would hold only half that amount. Eventually it was agreed that we could come back another day and pick up the rest. Finally, we drove in through the gate and round some tracks past huge buildings and in behind huge piles of lovely looking ground limestone.

I could write another page or two about the saga of filling a bag with more than a hundred kilograms of limestone and loading it into the ute but suffice to say many men were involved – all with an opinion (different) as to how it all should take place.

At 3.00pm we got away from the factory loaded up with my seven passengers and a bag of limestone in the back. Yes – limestone - to start adding to the diet of the dairy cows and calves to improve their calcium intake. We identified lack of calcium as a major issue months ago and at last we have slayed that dragon.

Finally, home at 3.30pm – starving and dying for a cup of tea.

Pip had spent a lot of the morning with Lami who is the three-year-old son of the farm cashier Kidist. Kidist lives in a small hut on the farm. Her husband comes to the farm occasionally but is not popular with the farm staff. When Kidist is working or is away from the farm, Lami is looked after by everyone at the farm, and he is spending more and more time at our hut. We are very much becoming his grandparents, and he is already speaking quite good English. – unfortunately, very much with an Australian accent!!! Often, he is waiting outside our door when we get up about 7.30am and he is with us all day, mostly helping Pip with her jobs or walking around the farm with either of us. Sometimes he drops off to sleep at the end of the day in our hut, and I carry him the few hundred meters back to his mother in her hut.

Pip getting some help from Lami - our constant companion.

This morning Pip also had the other youngster Hordofa (son of our shopkeeper Aberash) who is about ten and very clever at most things and has some reasonable English. He is a lovely young boy, and his drawings are the best art we have seen in Ethiopia – no offence to the Ethiopian Grand Masters. Pip had also been pulling up some old, barbed wire which man, woman and beast had been stepping over and through for years when walking up to the far end of the farm.

The two of us went for a farm walk late and picked up the collected barbed wire as we came home.

Some water appeared in the tap in our courtyard so obviously the village water system is working again. After a number of days without water, we were finally able to have a shower and wash our hair. Such a treat in a corrugated iron box with cold water trickling out of a pipe above!!! When the water system is down, we can have a wash under the irrigation bore, but that choice is pretty smelly with sulphur, and we always attract a crowd to watch us washing as there is no privacy under the water outlet from the bore.

Spaghetti bolognaises for dinner but every time I cook it, I dream of the beautiful Spag Bol that my brother-in-law Don makes in Australia, and my Ethiopian effort seems so feeble.

EE.26: The Tough times:

I suspect that anyone who has spent time in the developing world will know that there are stumbling blocks along every path, and we certainly had times when life got pretty tough. Fortunately, we had each other to rely on and when one was down, almost always the other would be seeing some glimmer of hope........

Day 207

There are times when Ethiopia is not our favourite place on earth and maybe tonight is one of those, particularly for Pip. We have had no power for 24 hours and no internet today. It is the coldest we have been, with heavy rain late this afternoon and I think that cooled everything down.

After the rain I braved a cold shower which was Ok and then Pip walked over to the shower hut, undressed, and stood freezing cold in the nuddy for 5 minutes with no water as it had stopped for unknown reasons. Sometimes you wait a while and it comes on, but nothing tonight so she had to redress and come back to the hut, not very pleased with things!!!!

With no power I ended up cooking mashed potato and heating up some left-over gravy (from last night's yummy roast lamb) and frying up a bit of the incredibly tough oxen meat that is the only option from the butcher. Not much to do sitting with our solar lights and no computer, so to bed early after Pip had her bath in the plastic bowl with a couple of litres of warm water heated in our kettle on the biogas stove.

One of my brothers has asked by email if the frustrations are getting harder to handle or are they so common that we get used to them. The answer from my perspective is both. We tend to move on quite quickly when blatantly obvious productivity issues are not attended to - on the grounds that the farm is now reaching perhaps 20% of its potential whereas the average for Ethiopia may be 10% (no idea of the real figures but that is the way it feels).

On the other hand, for the first few months I kept thinking that we are obviously not getting our messages across properly. Maybe it is the language, maybe we are not targeting the right person, maybe it needs to be simpler, maybe we need more detailed explanation, maybe it all needs to be written down, maybe in English, perhaps in Amharic, or perhaps I am not just used to working in a large organization? Every time the glaring, obviously needed change did not happen, we blamed ourselves and tried to change our technique. Now we get more frustrated because we cannot think of anything else to try.

EE.27: Law and Order:

Ethiopia has a formal judicial system, complete with a Supreme Court and a hierarchy of lower courts. But at the grassroots level, justice and conflict resolution often follow a more traditional path. In rural areas, it's usually the respected elders who step in to resolve disputes. Each kebele—the local administrative unit—has a designated "head" person. This role is voluntary and not necessarily tied to age, but rather to the individual's standing in the community. It's typically someone deeply trusted, often with long family ties to the area. Although Tibebu is widely respected, he wasn't born in this region. As such, he works closely with the kebele leader and other influential community members—mostly men, but occasionally women—to navigate local conflicts and other issues.

Day 212

A very quiet day as no farm staff were about except down around the cow shed. With Lechisa and Zinash away for a holiday, our courtyard is very quiet on the weekends with just the odd guard wandering about.

I made some hot bread for lunch which I am finally starting the get right – baked in our saucepan with a small sheet of corrugated iron in the bottom.

Pip had a walk around half the farm to check up on things and get some exercise. To our surprise Tibebu appeared late in the afternoon. He had been at a community meeting down the road most the day.

Apparently, there have been some cattle and sheep stolen in the area recently. This has not happened for a number of years, so the community decided to repeat the solution that they used effectively six years ago. They had a policeman come from Guder and today called a big community get together at the community hall. Firstly, there was lots of general discussion about what has been stolen, when, and the possibilities of how.

Everyone was then given a pencil and paper and told to write down any suspicions that they have, any circumstantial evidence and to nominate who they think may be the thief. Everyone then must put their paper into a box.

The policeman and elders then opened the box in private and looked for likely trends in evidence or suspicions. On this occasion, based on this investigation process, an older man and two young accomplices were taken by the police to Guder for further questioning. How do you like that for good grass roots policing???

We are not sure of the ethics, but Tibebu reckons it solved the problem last time and he expects it to have the same consequence this time.

Chapter 8

Sept 2014 - Worlds Apart

EE.28: Life in the fast(er) lane:

The Australian community in Ethiopia was small, and over time we came to know nearly everyone from Australia living in the country. From time to time, events were held at the ambassador's residence—known simply as The Residence—and most Australians, including those of us on volunteer assignments, were invited. The new ambassador arrived shortly after we did, which gave us the chance to get to know him and his family quite well. We also formed a strong friendship with the Embassy's consul, Stuart, and his wife Linda.

Day 217

We had decided that we would try and get away to Addis Ababa about 11.00am thinking the usual buggering around would take place and it would take most the morning for the ute to be loaded. By some miracle the ute was loaded and back outside our hut by 9.00am. Pip and I both had a few jobs to do, and I got the perches fitted into the chicken caravan.

We headed off after morning coffee, the ute loaded with two hundred litres of Niger seed cooking oil which had been processed at the farm, and forty kilograms each of butter and cheese. Due to the incredibly complex paper trail, for the first time I had to sign multiple copies of documents as 'receiver' of the produce and then later at the head office again as the 'dispatcher'!!!

Pip as always did a competent driving job to Addis, and we negotiated our way through the traffic despite very heavy rain. Some of the day was spent at the head office and other jobs around the city sourcing farm supplies.

Late in the day we went to the house of our fellow volunteer Gayle to freshen up and get into our "best clothes" ready for a reception at the Australian Ambassador's residence. We took a taxi to the house of some other AVI volunteers – Sasha and Peter a young Australian couple who have their two boys aged five and seven with them in Ethiopia. By now it was 8.00pm and Pip and I were starving as we had missed lunch again. We were worried that the meal may be over by the time we got to the Ambassador's residence as the reception had started at 7.30pm.

Fortunately, the taxi driver knew the way, and we were there quite quickly - to a most beautiful house and garden on the side of the mountains overlooking Addis Ababa. There was lots of security people about, but we got in OK to find speeches already underway. The reception was to farewell the current Ambassador (Lisa Filipetto) and there were about one hundred and fifty people present.

About half of the guests were from other embassies – including many Ambassadors, and the other half were most of the Australians either working or volunteering in Ethiopia. Speeches were soon over then a most beautiful meal presented by the Hilton Hotel catering department with endless drinks and chat.

Pip and I knew some of the Australian embassy senior staff from when they had come to the farm, and they seemed to be particularly interested in our activities and made quite a fuss that we had come in from the country for the evening. We met a heap of interesting people including Colonel Wes who is the Australian defence attaché for the whole of Africa. We also met a lovely Australian guy Steve who manages the Born Free game park/reserve we had passed many times on the road from Ambo into Addis. We discussed a visit to see the park when we returned to the farm on Sunday.

The music cranked up, and we drank, danced, and chatted until 1.30am in the most beautiful setting imaginable. There were even lamingtons and Anzac biscuits on which to "graze" and I think Pip and I ate enough delicious food to last us a week. Some good quality red wine slipped down very easily!!!

As the night wore on it was mostly young Aussie volunteers (including us!!) grooving away when eventually the ambassador was suggesting perhaps enough was enough. During the evening she had worked her way around everyone for a good chat, and made us all feel really welcome.

Heading home, six of us wandered out the gate and down the hill towards the main road hoping that we may be able to find a taxi (very little chance we later realized!!!). Fortunately, out of the ambassadorial residence car park came one big Land Cruiser with Luxemburg flag fluttering on the front and then another with the Georgia flag. They passed us then stopped, and a lovely girl from the Luxemburg embassy with whom we had been dancing (and Pip thought looked exactly like Audrey Hepburn) hopped out and suggested we may need a lift to somewhere - We sure did indeed!!. She flagged down the other cruiser, and it had a driver and the Georgian ambassador inside. He was also young and had been grooving around the dance floor with us. (Pip thought he looked like Prince Frederick!!!).

Fellow AVI volunteer Peter, Pip and I hopped in with the Georgian ambassador and his driver and at 2.00am, we dropped off the driver at his (shanty area) home and then the Ambassador drove us across Addis and to the door of Peter and his wife Sasha's house where we were staying for the night. Sasha, Gayle and another Aussie, Amy, had been dropped off a little before - at the Luxemburg taxpayer's expense. Such a lucky break to get home!!!!

EE.29: More Frustration with Bureaucracy:

Some days on assignment were more comfortable than others, but it was rare for a day to pass without something extraordinary redirecting our lives. I imagine an Ethiopian visiting our farm in Australia might find a few things surprising—but not the kind of daily, jaw-dropping moments we seemed to encounter. We couldn't help ourselves—we had to record them.

Day 231

Back in Addis Ababa for an excursion to try and finalize the purchase of chickens for our newly constructed chicken caravan. We were up early in the hotel for a second hot shower making the most of the hotel room and then used the good internet connection to Skype family in Australia.

We expected Tibebu and a driver to arrive in the farm ute any time from 8.30am as they were supposed to leave the head office at 8.00. Pip was swearing and cursing about the lack of exercise being locked in a hotel room, so she headed off to the gym whilst I waited in the lobby looking on the internet for flights to London for Christmas.

Tibebu eventually arrived at 10.00am so I went and collected Pip (who no doubt had bench pressed 300kg in her frustrations and run 25 kms on the machine) and we piled into the ute which also included the purchasing officer from the head office as she was bringing the money to pay for the chickens and the feed we hoped to purchase.

It was clearly going to be quite a long day with five of us jammed in the ute with Pip and my luggage on our knees. We cannot leave stuff in the back of the ute, or it will be nicked at the first traffic jam. I was going to say nicked at a "set of traffic lights" but there are only about three sets in the whole of Addis, and we were unlikely to be near any of them. We were not sure where the day was going to end up, so took our luggage with us.

It was a drive of about one and a half hours to get to Debra Zeit to look at the farm where we were to purchase the 3-month old chickens. Also, the manager wanted to talk to us about the raising of chickens.

We expected the drive to be about forty-five minutes as there is a new highway just been opened down south of Addis Ababa but for unknown reasons our driver took the old road. Tibebu said it was because on the new road everyone goes too fast, and the cars keep “jumping the fence”. The mind boggles!!! We think it was probably because our driver (supplied by head office) did not know how to get onto the new road.

Our frustrations were compounded when we got pulled up by the police and issued with a fine of three hundred birr ($15) for a cracked windscreen. At this point the fine is issued, and the police officer takes the license of the driver and casually stuffs it into the top pocket of his shirt. The fine would not have been issued if we had been prepared to pay the usual bribe of two hundred birr, but I don’t think Tibebu, or the driver like to do that in front of us clean living volunteers!!!

So begins a long saga, but before describing that, we did get to the chicken farm (just at the moment of torrential rain) and were ushered into an office where the purchasing officer paid the nine thousand six hundred birr ($A480) for one hundred chickens. We waited for about half an hour for a manager to come and he was a great guy who took us to his office and in good English gave us a quick lesson on the raising of chickens and, of more particular interest to us, the likely diseases in Ethiopia and what vaccines they will be given etc.

We were pleased to find they sell transportation crates that will fit into the ute to transport one hundred chickens back to the farm in a couple of weeks. As we have some money left from our grant, we have decided to purchase the crates so they can be used in the future to pick up more chickens.

We then headed to see the shop which the chicken farm owns a couple of kilometres away, but the time being 1.30pm it was closed as they have lunch “hour” from 12 until 2pm. Anyway, the chickens will be right to pick up on October 25th and on that day, we can buy boxes, egg trays and any other bits and pieces that we need.

Next stop?? We set off for the road traffic authority building in the town area close to where the police had stopped us this morning to pay the fine. Through lanes and back streets (asking direction of locals all the time) to a large fence with a gate. No sign, no nothing. In went the driver to pay the fine with our “donation” of 300 birr and out again after 20 minutes or so. “Ah the traffic authority cannot issue a receipt today – Friday. Come back on Monday”. For this the cynical should read “pay today, we will write a note to say that you have paid, you will be able to take that to the policeman to get your license back. We will pocket the cash; there will be no receipt in the book – everyone happy”. Tibebu sent the driver back in with the 300 birr and we got the required note on a piece of torn paper with a stamp on it.

Off to the police station. “Ah –the police officer that booked you has been back to the office but did not leave your license and now has gone home for lunch”. It was now 2.20pm so that seemed most unlikely. We were sent out of town to try and find his house, but no luck, so returned to the police compound with our driver very keen not to return to Addis Ababa without his license. This time, after another half hour wait, the police said they would lead us to the officer’s house. Ironically having been booked for a cracked windscreen in our ute, the old Toyota Hilux ute that was the police car had no back lights operating and was a complete wreck.

After a time, our accompanying policeman decided that the cop with the license had left his mobile phone at home, but he was not there himself (there were suggestions he was perhaps visiting another wife???). Eventually, as we were heading back yet again to the police station, we spotted the officer walking on the side of the road apparently having visited his afternoon wife. We pulled up, showed him the hand written note saying we had paid the fine and he pulled the licence out of his top pocket, and off we went – with one Australian male in the ute very keen to give the offending cop a full double barrel burst of abuse and recommendations for his future, but by then our driver was at full speed heading for Addis- never to look back I suspect.

Next stop was the stock feed factory who had won the tender to supply our chicken feed. The plan was to pay them for the feed today, ready to also pick up on the 25th of October. On the tender document they had written Kality Stock Feed Processing Company - so that is what our cheque had been written out to by the finance department at head office. Unfortunately, it was old letterhead, and the company name is now Kality Stock Feed Company. They could not possibly accept our cheque (and our purchasing officer could not possibly change the cheque) so off we go. Hopefully we can pay when we pick up the feed...maybe, maybe!!!

EE.30: Rabies – safety and more moral dilemmas:

Day 239

At the farm, the labourers have been re-ploughing the potato patch for the third time and still finding spuds which they are piling up in the hayshed. We had suggested to Tibebu that they may broadcast some oats onto the potato ground, and we will disc them in whilst there is still some moisture. Unfortunately, it did not happen today and with tomorrow Sunday there will be no action until Monday so the moisture will probably be gone. Note to management – TIMELINESS!!!

Sorting potatoes in the hayshed.

Tibebu spent the full day at another large community meeting where the thieves that had been identified by the "community" circumstantial evidence were made to stand in front of the large crowd who again presented their evidence, and the thieves eventually confessed to all the thefts that they were suspected of. Now they will show up at another meeting where the punishment and repatriation will be decided.

Pip washed towels and did a fair bit of work in her veggie gardens which are constantly expanding. The locals are amazed at all the different things she grows and cannot believe that if you have grown corn or beans or something once this year that you would just grow it again in the same year. Surely if you have grown beans once this year you can rest from having to grow them again!!!

Pip went to the internet spot and checked up about rabies, as in the middle of the afternoon we sensed that something had happened in the community. Young twelve-year-old, Hordofa, son of the farm shopkeeper, was with her. He is now our main link to the world as he is the only person around who speaks good English when Tibebu is not here. Pip asked him what had happened. It turned out that a daughter of one of our guards and her friend had both been bitten by a rabid dog. Tibebu later told us that some people had left the community meeting to start running after the dog. We did not hear if it was killed but most likely. Apparently, everyone goes onto high alert now as it is not known how many other dogs may have the rabies. Even the guards and shepherds need to be alert that none of the calves, cows or oxen get bitten as they will also die.

The two children waited with their families on the side of the road for a few hours for a bus to head down to the clinic at Guder which has the vaccine/treatment. How would you like to do that? Waiting for an unreliable bus, knowing that if you did not get the required treatment with a certain number of hours, the virus gets into your spine and brain, so you will be paralysed and soon after die? Apparently, the odd time we see an ambulance on the road it is for serious injury or women in labour – not little problems like rabies!!! We are not sure why someone did not come to the farm to get Pip or me to take them down to the health clinic in the ute. Probably because the children were "only" the progeny of a guard and not a manager!! Makes me cry just thinking about it. Pip and I will be making sure we walk with a stick for the next while and will avoid the road where most of the dogs seem to congregate.

We understand that for rabies treatment, the children now receive an injection into their bellies every day for 17 days.

Chapter 9

Oct 2014 - Birbirsa Life

EE.31: The ups and downs of everyday life

Pip and I have been married for thirty-three years, but nothing truly prepares you for being side by side, every hour of every day, in a place so far from everything familiar. For over eight months now, we've rarely been more than a few steps apart, leaning on each other completely, not just for the practical things, but for strength, reassurance, and understanding. It's tested us in ways we never imagined, but it's also revealed something mightily important on an assignment like this. The small tensions and differences that once might have lingered for hours—or days—now seem to dissolve in minutes. There's an unspoken but necessary rhythm between us now. Maybe it's survival. Maybe it's love. Probably both.

Day 243

Pip headed off for a walk with the binoculars before breakfast and, as usual, added another bird to her list.

She headed off again in the late afternoon to measure a few eucalypts with a diameter tape she had made. Again, with the binoculars in hand she saw a lot more birds, including a couple of incredibly colourful but elusive White-cheeked Turacos. (These are Pips notes)

"They are large, 45 cm, birds, with a bright green front and head, bright blue wings and tail, with a blue topknot, bright red beak and eye ring, and small very white patch in front of the eye and a white streak behind eye, and in flight they have brilliant red on the tops of their blue wings."

She continues to add lots of bird species to her list – many unique to Ethiopia or East Africa.

By late in the day, I had a very sore back, there was no power, a flat computer, no water for us or the cattle, and I was feeling about as flat as the computer battery. I had hoped to be able to charge the computer from the cigarette lighter in the ute, but I must be too dumb to work out the correct leads or something. My whiskey ran out a week or so ago and we are out of our comfort food – Pringles which we keep for "pick me ups."

But then I lit the campfire, put in a couple of huge freshly dug spuds in foil, and we sat and cooked lamb chops under a beautiful African sky with not a light to be seen anywhere - and life did not seem too bad after all!!! By candlelight (using a beautiful, coloured glass candle holder which throws out many colours, given to us by Ali and Sophie when they arrived to visit us having travelled from Turkey) we climbed into our comfortable bed where we invariably sleep like logs for 8 hours or more, ready to start a new day refreshed with enthusiasm.

EE.32: Diet

Day 246

Before breakfast I had a local farmer at the door wanting his axe sharpened. Now that I have the angle grinder this is no problem when the power is on. As the noise from the grinding began to echo around the community, I suddenly had half a dozen farm hoes to sharpen, and a number of axes also appeared. It was quite a while before I got back to finish my breakfast. I think a lot of the local farmers hoes and axes that I sharpen had been thrown into the back of a hut years ago when they had become completely blunt and useless. Now the farmers hear about the "farangi" with the loud, sparky machine that can bring the old tools back to life, they line up pretty regularly whilst I reshape their hunks of old metal. I suspect that often I make them better than new - when they were probably previously just sharpened by bashing them into shape.

Being a weekend, we had headed to Ambo and had a large lunch with friends at the Abebech Hotel. For dinner back at the farm, we did not need much to eat so lit the fire and roasted some maize which one of our guards had given us. We have talked to Tibebu a lot about the diet of the local people, and it must surely be one of the biggest differences between our western ways and the community here. I am sure it would not be exaggeration to say that most – if not all - the locals here would eat only about ten to fifteen different foodstuffs in a year. In Australian that figure would surely be close to 1000, maybe more. The rural Ethiopian people all eat teff which is a fine cereal grain, ground into flour, mixed with water and fermented for a few days and then used to make the injera (pancake). This is eaten for EVERY meal. Sometimes if the teff supply has run out or cannot be afforded, they will grind maize into flour and use that for the injera. Maize is also roasted on hot coals and eaten straight off the cob. Vegetables regularly eaten are onions, garlic and chilli which are used in almost every wot (stew). Some of the wots will also include cabbage and tomatoes and occasionally potato. Beef or chicken may be added to the wot but mostly only in festival or celebration time (funerals!!!), and raw meat is often eaten during festivals. Shiro made from lentils is a thick soup like wot, usually spiced with chilli powder and is also eaten regularly. Some people would eat eggs and drink milk and at times eat dry roasted barley. The local "false banana" called ensat is also used to make bread, particularly in the highland areas. According to Tibebu, even if we gave them beetroot, carrots, lettuce, beans, peas from the garden or any of the other processed or packaged food on our shelves, they would not eat it. Certainly, they have a simple diet. Not a lot of decisions to be made as long there is plenty of chilli powder available!!

EE.33: Helping the local community

Pip and I continued to be completely absorbed with our assignment of turning around the fortunes of the Abebech Gobena farm at Birbirsa but along the way we were delighted to be involved with the whole community in other activities and help out where ever we could.

Day 254

Whilst we were having our breakfast (with little Lami sitting up on his stool sharing our oranges and porridge as usual) Hordofa and his older brother who is a school teacher in Ambo, appeared at our hut. They told us that their cousin had purchased a small new diesel water pump and was intending to water some newly planted sugar cane out of the river which runs between us and the village of Dogoma. They had tried to get it going yesterday but could not get it to pump water so maybe the farangi's could help?

We knew the approximate location, which was quite a few kilometres away, but we were happy to walk. The four of us set off with a few of my tools and a bucket, as I suspected they had not realised that pump would need to be primed from the river. Despite not being our favourite place to walk as it is a flat gravel track only used by local people and donkeys with no trees or shade, the morning light on the hills and mountains in the distance, and all the crops just in head and full flower, it was quite stunning scenery. After maybe a five-kilometre walk, we got to the paddock. There was a fairly steep climb down a path to the edge of the river, and we were greeted by half a dozen men and boys digging, hacking and clearing a line for the "lay flat" 4-inch canvas hose from the pump near the river up quite a distance to the sugar cane field. I was a bit nervous as it looked like a pretty big pressure head to me, and I was doubtful if the pump could do it.

First, I wanted to see if the motor would run and after a time, we got it running. A neat little Chinese diesel but felt and sounded pretty tinny, so I don't know how long it will last. We carefully connected the suction pipe ensuring that the connecting flanges were properly seated (maybe one of the problems they had yesterday) and with plenty of patience and priming, it started pumping a huge flow of water. Lots of handshakes, clapping and general excitement. We turned it off and connected the delivery pipe whilst I took the opportunity to give a lecture on "head" and "lift" and "friction in pipe" and the connection between them and the amount of fuel the pump will use. It was all translated from English to Afar Oromo by the big brother, and seemed to be well absorbed by the crowd standing around.

Advising on the installation of a new water pump

We fired up the motor again and despite my trepidation it pumped a good flow of water up to the sugar cane and irrigation began immediately. We gathered up the tools and walked back up to the top of the riverbank and big brother invited us to his family house/hut for bunna (coffee).

We walked across the lush countryside to his house and there we met lots of gorgeous looking children and a few more adults and sat inside the hut with the mud floor. We could hear the coffee beans being roasted, then pounded in the room next door through the customary curtain/drape over the doorway. As always, the whole thing took the best part of an hour and by then the wife and daughters were starting to bring out the injera and wots. Sadly, we had not brought our water bottles, and we were aware that the only water would be from the river (which we do not drink without boiling), it was already a warm day, and we had a long way to walk home in the sun. We reluctantly declined the injera and the wots which looked lovely. They kept assuring us that they were not spicy – figuring that was the reason we would not share their meal. Eventually the bunna (coffee) was made and Pip and I both had a couple of cups to try and get as much fluid into us as possible. By then the others were finished their injera, so we set off for the walk home - with the usual hundreds of people en route to greet and acknowledge and handshake and practice our Afar Oromo greetings, and a few children to hold our hands, donkeys to avoid and all that goes with a walk in the community. Home about lunchtime in pretty good condition, given the morning's adventures.

In the afternoon, Pip did more washing and was soon besieged by Lami and Bargar the son of one of the guards (both of whom were here until dark.) Then came an older son of guard Abara, who brought his grade 6 English book to read to Pip. He was followed by three girls from the village who also got involved with reading books to Pip and practising English. Somehow Pip also managed to make biscuits and a chocolate cake along the way.

EE.34: Management – Another visit from Head Office

During the twenty-five years that the organization has "owned" the farm it has been a passion of the founder Abebech. It was she who appointed Tibebu as the first manager. He had retired for a brief period about ten years ago but had returned as manager of the farm after a couple of years away. Tibebu and Abebech had a very good relationship – enhanced by the fact that every one or two months Abebech would travel to the farm for an overnight stay to check on proceedings. She had a strong environmental interest and as the farm had developed, she had insisted on retaining the patch of jungle/native forest along the creek that ran right though the centre of the farm. She also was determined to maintain the single paddock trees and patches of scrub scattered around the farm. In her declining years Abebech was not able to come to the farm but the general managers came every few months and the board of the organization came once a year to view the activities.

Day 256

Another perfect morning and everything looking splendid for the visit of the managers and all the board members of the AGOHELMA organization. I got up early, as I was pretty sure the farm management would want Pip or me to drive them to Ambo early to pick up the meat to be cooked today for the visitors. Sure enough, they were all ready to go at 8.00am. On the way to Ambo, we had with us the father of one of the children that was bitten by the rabid dog a couple of weeks ago. With the help of Tibebu interpreting, I found out that the two children who were bitten are OK, but they were up to day 14 of 17 consecutive days that they must go to Guder and get an injection (through the navel by the look of the sign language). Sounds incredible but apparently that is the way it is. Hopefully all will be well with them. I had seen the boy a couple of days ago and he has a horrible wound on the back of his calf muscle. It is healing OK but must have been incredibly painful.

Getun drove to town quite capably under my guidance with only one major scare. We very nearly ran over 2 men who were walking on the side of the road and a bus came around a corner towards us a bit towards the centre of the road. This called for Getun to share the same patch of gravel as the pedestrians. I suspect they have their underwear on the clothesline tonight!!!

We got back to the farm safely with the meat to be cooked up for lunch. Soon after, the head office vehicles arrived. The two senior managers, five members of the board (one of whom was one of the two orphans that Abebech Gobena took in her care in 1980 when she founded the orphanage) plus the secretary to the board, the lady who looks after the farm interests in head office, a lady from finance and fellow AVI volunteer Gayle who is now in charge of public relations at the organization.

We all went straight into Tibebu's office, and he gave a presentation of the farm activities, and then I was invited to present what we had been doing and our plans for the future. It all seemed to go well.

We all then went for quite a long farm walk, looking at the irrigation bore (no electricity today, so could not pump water for them), our climbing bean trial, the irrigated tomatoes, magnificent looking teff, niger seed and oat crops, back to the chicken caravan, onto the milk processing and biogas, oil extraction, the dairy cows and the new cattle crush set up ready to go. Pip and I, the farm staff and the farm management team felt we had plenty to be proud of.

Pip with Abebech at our experimental plot of climbing beans on maize stalks.

Back at our courtyard the female staff had prepared some yummy un-spicey wot with all the freshly purchased meat and injera and bread, so we sat around and had an enjoyable lunch washed down with glasses of milk - or for some of us glasses of Tej which had been brought out from Addis. Also of course, the customary bunna (coffee). Pip and I had a good chance for a catch up with Gayle.

Then we were called together again by the chairman of the board and Tibebu spent the next half hour praising Pip and I for our efforts both on the farm and in the community. One would think we are changing the course of Ethiopian history with all the nice things that were said. Hopefully something might be different for some people in ten years' time.

We were also asked to raise our concerns, so I went in pretty hard on the difficulty of sending large amounts of profits to Addis whilst we have so many overheads to sustain on the farm – principally all the guards and managers. I pointed out that we would just have to get more productive and have everyone working more if they wanted to keep their jobs. Pip raised the problems of replacing Lechisa who has recently resigned and taken up a much better paid job on a coffee plantation. Also, Pip pressed the need for a young new bull on the farm which is becoming a major issue for us.

They all left at 3.00pm and Tibebu and the other managers sat around the rest of the day chatting excitedly about a very successful day.

EE.35: Opportunity – diversity of human nature

It's often only when we step away from the routines of everyday life that we truly begin to appreciate the remarkable diversity of the human race.

Day 270

Melbourne Cup run and won by a German horse we have read on the internet. No one here will have heard of it of course. In fact, we have not yet met one person here in Guder who has heard of the Beatles or Abba!!

Late in the afternoon we went for a walk up the road to see a paddock of red teff I had noticed a few days ago and I thought it would look stunning in the late afternoon sun. Not quite as vibrant as a week ago but we still took some great photos and we also came across a boy shepherd who was drawing whilst looking after his cows. He had some decent Texta's and an art book, and for a boy of about twelve, he was a very good artist. I feel a bit ashamed to admit that when we came to the area, it felt as though everyone was pretty homogenous, but then as you get to know people more, you realize that there is the same diversity as we find in our Australian society. The lack of resources often mean that the artists mostly don't have the materials, the sportsmen no balls, the musicians no instruments, the actors no opportunity to act and I suspect the academics no opportunity until later in life. Plus, all the children from about 3 years of age have numerous responsibilities - making coffee, carrying water, shepherding, cutting and carting grass and crops, cooking, weeding crops, chopping maize, collecting firewood etc. Our young mate Hordofa must be the worst shepherd in Birbirsa. So often he is here on our computer, reading to Pip, drawing, or playing his whistle/flute and suddenly "oh no, my cows" as someone from a kilometre away will have yelled and reminded him that he is supposed to be looking after cows. Other boys will be attentive to their cows every minute of every day. Three-year-old Lami is just completely fascinated with anything mechanical but if we were not here, the closest he would ever get to something mechanical would be an axe or shovel. He loves my pliers, shifter and other tools and is always in, under or on the ute, if I am fixing something.

Clearly research tells us that a certain proportion of every population is homosexual. It would be fascinating to know if that diversity is also in this community. As homosexuality is unlawful and it certainly is not spoken about, there is not the slightest hint of what proportion of the community are forced to live a tormented or unhappy life as a result. Ironically, it is commonplace for men (and occasionally women) of all ages to walk along holding hands.

As we came back from our walk, we found the neighbour had 30 goats and 25 cattle grazing our land down in the far corner of the farm where they could not be seen. We had come in the back way and surprised the shepherd. I took a couple of photos to show Tibebu. As usual we had three guards sitting around the farmyard area doing nothing. The neighbours are also aware that our staff all knock off at 5.00pm (11.00pm Ethiopian time) and about then our guards accumulate around the farmyard until dark. It is then open slather for the neighbours to get in some good grazing. Pip and I are the only ones out and about at that time.

EE.36: Local women's group – justice system

Ethiopia has long been the focus of countless international aid initiatives. In rural areas, these projects can achieve lasting impact—especially when paired with strong local involvement. In this instance, an Oxfam project was brought to life through the efforts of a local women's group, which has since become a vital and deeply respected part of the community.

Day 277

I did some washing in water heated up on the biogas and Pip headed off to check cows and calves. We did a few more jobs around the hut and just when I was hanging out for a morning coffee, Tibebu came and found me and said he wanted to go immediately up to Mutulu – the village seven kilometres up the road - to stop the water coming into the irrigation channel.

Just outside our farm yesterday afternoon the grader was grading our road (a miracle), but he went off into the table drain which was filled with water by an overflow from the irrigation channel running down from Mutulu. Now, graders are really hard things to bog, but once bogged they are really, really bogged and very, very hard to un-bog. By the time the driver had fiddled around most the afternoon, this one was up to its belly in soup with all six huge wheels almost out of sight. Huge dump trucks are currently taking rock up the road to somewhere in the highlands, and the grader driver must have convinced one to stop and dump a huge load of rock on the road beside his grader. There it sat for the night.

Hence the need for our trip to Mutulu to turn off the water so that the area may eventually dry out, and the grader can be retrieved. It didn't sound like a big ask and I was sure that I could survive half an hour without any sustenance. Fool, fool, fool. How could I fall for it after 9 months?

A man who had obviously never been in a vehicle "climbed" into the back seat (they never slide in like the rest of us) and off we went – me thinking he was something to do with the water. Tibebu then started to tell me that the fugitive in the back seat was the cause of some serious angst in the community, and we were taking him to a committee meeting to work out his guilt and sentencing. Suddenly my hunger pains were thinking – this is just not a morning tea missed but a lunch as well and maybe more. So much for just turning off the water!!

It turns out that a few years ago the local women's group (founded by the hard work of our farm manager Tibebu) got funding from Oxfam to purchase 120 heifers (young female milking cows) which would be given to women to give them a start in life. Tibebu was on the purchasing committee. There were 300 women in the group so it was agreed that there would be a ballot for the heifers with the first 120 women drawn to receive a heifer. Then, any female calves that were born to those original ones would continue to be distributed to the other women until all 300 women had one cow to milk as they matured. Any male calves could be kept by the family that owned the heifer and obviously after the heifer matured and had her own calf, they would have the milk to feed their families or to sell.

The wife of the little man now huddled in my back seat had received an original heifer, but it had been discovered that it had since had 2 heifer calves that he had sold at the market instead of handing them on. Time for the committee to meet and sort out retribution and repatriation!!

Having made up my mind that I could be sitting in the ute on the side of the road on the outskirts of the Mutulu village with nothing to do for most the day I determined to teach some English to the children that had started to accumulate around the windows of the ute whilst I waited for the "kangaroo court" to deliberate.

It was only about an hour before Tibebu came back to the ute and the little man headed off walking down the road. His guilt had apparently been proved, and sentencing would be at a later date by the Kebele (local government). He will not be taken to the court in Ambo which apparently pleased him mightily.

Meanwhile Tibebu had organized someone to turn off the irrigation water, but as we were driving back to the farm, the grader appeared driving up the road. The grader operator found that he could lift himself up by pushing his rippers and blade down and then he conscripted dozens of kids and adults that had been standing around since yesterday to throw all the rock from the road under his wheels.

I was happy to be home and have some lunch.

Chapter 10

Dec 2014 - Holidays and Happiness

EE.37: Holidays – Ethiopian Tourism

Australian Volunteers International (AVI) had a small number of volunteers in Ethiopia and Tanzania and each year we all got together for a few days of workshops and socializing. In our first year of assignment, it was held in Hawassa, a city three hundred kilometres south of Addis Ababa. After the event, Pip and I stayed on for a few days holiday and then had the unforgettable experience of a long-haul bus trip back to Addis.

Day 296

Our phone alarms had us out of bed at the Hawassa Haille resort at 5.15am and despite some doubt as to last night's promises from reception that we would be able to get some breakfast at 5.30, kitchen staff did appear, and we even had some nice porridge.

They had organized their courtesy bus for 6.00am to take us to the Salem bus station where our big bus was to depart for Addis at 6.30am. The Salem buses are forty-seaters of varying condition. Some of the ones we see going through Ambo are old and decrepit.

This one was quite modern, and our large bag went in the locker underneath whilst we had front row seats – good for watching the traffic/driving???? Whilst sitting waiting to leave, Pip watched out the window at the pick pocket boys doing their thing. There were about twenty people crowded around the door of the bus waiting to board, and two boys about ten years old, looking like true ruffians pushing in amongst them. It is so incredibly obvious what they are up to as phones and wallets got relocated. Clearly, they were not about to get on the bus, but nobody does anything about it. We could not tell the extent of their reward this morning, but there were some older men standing nearby who were almost certainly waiting for the spoils which no doubt they would sell in due course.

The bus trip itself took about four hours. We had not been given any idea how long it may take. Best estimate we could get was a few hours to a day or so, hence we had taken some water and a few bread rolls from the breakfast counter at the resort. We also did not really have any idea where the bus terminated but assumed it would be in Addis somewhere. To our surprise after about an hour the "conductor" handed out a croissant and bottle of water to everyone. At about three hours the bus pulled up on the side of the road near a couple of low shrubs, and this was the toilet stop. Men just stood around anywhere peeing and some of the women just squatted in the open, others vaguely concealed behind the shrubs. OK if you have a long dress and no undies like most of the local women - a bit more exposed if you are wearing pants. Luckily, Pip did not need to go! Apart from private chartered vehicles this is the only way for tourists to get to Hawassa (there is no airport yet as far as we know) so it becomes clear that the tourist industry has a very long way to go, as I am sure not too many international travellers would have found this pit stop very satisfying.

There is no point me trying to describe the driving. A true case of "seeing is believing". Suffice to say that the drivers of these buses think they are the most important people in the world and the road was built exclusively for them to manoeuvre a large bus as quickly as possible from point A to point B. Apart from the hour or so when we were on the new freeway, the horn was used constantly. The longest period I counted between blasts of this klaxon was 13 seconds. Absolutely no consideration is given to oncoming traffic (even fuel tankers) or traffic to be passed – just a blast of the klax and up the middle of the road we go. Moving to left or right was done as if the vehicle was a Hyundai I30 so the whole bus swayed alarmingly but somehow fitted into the tiny gaps that opened as a result of the endless horn blasts. Actually, all of this was not as concerning as the section of the freeway. Ethiopian drivers have no experience whatsoever of driving in uncrowded, relaxed conditions and many accidents occur as a result. This driver became more and more erratic as it appeared he just could not keep his eyes open. We wandered aimlessly all over the road, and he started jamming his foot up and down on the accelerator. All I could do was to wave my long leg and booted foot past his head every minute or two from my seat behind, which seemed to shake him into action for the next few minutes. We were very happy to see the toll gates appear and get into some serious Addis traffic which he negotiated with precision.

The bus terminated at Meskel Square, one of the biggest intersections/bus stops/meeting places in the city. This was convenient for us as it is not far from the house of our Australian friend Patrick. We walked to the nearby café beside the Red Terror Museum and rang Patrick who was at work despite it being Sunday. He was busy for an hour, so we had coffee, and he picked us up about midday. He took us back to his house and left us there while he went back to work.

Later we walked up to the Bole Road, Shoa supermarket. One thing all expats in Addis Ababa know is the location of the half dozen or so supermarkets (not bad for a city of 5 million). Every expat you meet can tell you which supermarket sells what, where the white porridge oats will be found, which aisle for custard powder, how much for the ketchup and where you used to be able to get so-and-so but it ran out 3 months ago!!! Considering how hard it is to find much beyond basic supermarket stuff in this city—and the fact that supermarkets are the only places where prices are actually marked on products, sparing you the usual 300% markup for being a farangi—there's real satisfaction in a successful supermarket run. Among volunteers, it's always a topic of conversation.

We had nearly finished filling our trolley with our monthly shop when we ran into Patrick who had popped in to get a few things to make our dinner. Most convenient as it saved us a long walk back to his house heavily laden with all our goods.

EE.38: Local Wedding in Highlands

Our farm is in a lovely countryside of rolling hills and valleys and is just a little higher altitude than Addis. While quite high at 2600 meters, we did not feel as though we were in the highlands. If you continue further up the gravel road past the farm to the village of Mutulu, the road really starts to climb. It does not necessarily feel mountainous, but after another twenty kilometres, you are seriously high at 3,500 meters. A wedding at a remote village in those highlands was inevitably going to be worth documenting.

Day 303

Another day of the wonderful Ethiopian episodes that enrich our lives and our assignment and washes away all the negative thoughts one ends up with after a stay in Addis.

On Friday, Dr. Fikru—the Ethiopian guy who first helped us and took us to build his father's house in the highlands—called us. As is his way, he "demanded" that today we would join him again at his father's house in the remote village way up in the highlands for the celebration of his sister's wedding and the blessing of the completion of the house. He was delighted when I said we could drive ourselves up as he was already bringing a carload from Addis.

Pip and I were about to drive out our farm gate when Carson, one of the farm guards, heard that we were driving to Dr Fikru's village in the highlands, and to our surprise he hopped in the backseat of the ute. We assumed he had friends or family living up that way and was keen to visit. He has no English at all but later when we had Fikru to interpret, it turned out that he was worried we may get lost or into some problems, so he just came to look after us- our own personal bodyguard!

After turning off the main road to Shenen, onto a very rough track we came across a man walking alone but dressed in a suit, so we stopped and picked him up. Turned out he was a brother of Fikru. Pip and I were confident we could find our way, but the one time we had been up there previously the road was so steep we had walked one section and had taken a short cut so had not actually been on all of the road. Today we would have made a mistake if we had not had the brother with us - lucky. The road is very steep and washed away so I had the Hilux in low 4WD to climb up the worst sections.

Eventually we got to the gorgeous highland village of Tulla Malke and were immediately surrounded by family members we had met on the last visit. The house which we had helped build months ago was finished, with a large main room and the very impressive large beam that we had helped carry from the forest stretching right across the building. Fikru was his usual organizing self – making sure we were well looked after and introduced us to everyone whilst also organizing everything else. It was his first trip to the house and to see his family since we were there months ago and the first visit ever for his wife, fourteen-year-old daughter and twelve-year-old son. Along with other members of the family from Addis Ababa, they had all walked in from the main road, as the track was virtually impassable – except for out trusty Hilux. His wife who is seriously overweight had been put on a horse and men had walked along beside holding onto her legs.

We were served "breakfast" a large pile of tibbs (lamb or goat meat) which Fikru had organized specially for us to have no spice and bread rolls instead of injera. Pip and I floated between inside and outside, including a walk down through the family's false banana plantation with one of the brothers who is at university – one of many of the brothers and sisters that Fikru supports financially with their education.

Now to describe Fikru's family - as told to me by Fikru: His Father, who is now 92 and has a bit of dementia but still a very interesting old man married Fikru's mother and bore 4 children. She then died before she was 30 from "disease". Father then married another woman and had 11 children with her. Whilst he was busy expanding his family with her, he did the normal thing and took on another wife whose husband had gone off to war and, after not being heard of for some years, it was assumed he had been killed. Under this circumstance another man is allowed to take her on and they subsequently had three children.

The unforeseen circumstance was that after 20 years the second wife's husband reappeared. This causes a bit of a problem as under traditional law he has the right to kill the man who has taken on his wife – given that he was still alive. Not a good scenario when father has by then fathered and taken responsibility for 18 children!! Apparently, it was all pretty touch and go for a while and Fikru was old enough by then to be involved in negotiations. Eventually the returned solider agreed not to do the killing but took back his wife and the three children and has looked after them as his own.

Around 11.00am we all headed inside the big single roomed house with about fifty adults sitting around on lovely long wooden seats made from local slabs with woven strips of cowhide for the seat area. Masses of kids sat in the end of the room. Fikru had each family group sitting on a bench and a large woven mat was spread out on the floor in front of the closest family.

It was clear to us that the wedding ceremony was about to begin. With lots of clapping from us all, in came the bride and groom-to-be – he in a resplendent white suit with small green Ethiopian trees dotted over it, and with a red tie and indescribable white shoes almost like Dutch clogs. (Nobody would even choose them out of the dress up box). The bride had a white wedding dress also with the small green trees all over it and with thin shoulder straps but slightly marred by two huge bra straps across her back – one beige the other light green. The need for two bras – of different colours with highly visible straps was not apparent to us! She also wore a cultural headset with shiny bits that hung down her forehead to her nose. The girl is a sister of Fikru and has been living in Saudi Arabia for the last few years and the Ethiopian guy had been living in the US for 10 years. He looked in complete shock the whole day, and even though he was Ethiopian born, we suspect had way less of an idea than Pip and me about the ways and customs of Ethiopian rural life.

The couple-to-be then proceeded to go around all the adults including us. First time around the room each of them gave us a small piece of grass then kissed us on both our shoulders then both of our knees. Second time around they each had a bottle of perfume, and each put a spray on our shoulders to check that we would accept them coming together with their own odours (yes really). They each had a helper (bridesmaid and groomsman?) who kept things in order. Both were gorgeous looking young people beautifully dressed.

The four of them then sat in front of the most important family benches and lots of short speeches were made by father, mother, uncles and at least one brother. Sometimes a bit of chanting accompanied the speech. All the time a young guy walked around with a large laptop computer presumably with a camera taking an endless video. It seemed odd to have an open computer stuck into your face periodically and in such a traditional setting far from the age of technology. Some of the younger men then started dancing and chanting and led us all outside where they continued the dancing, joined to our surprise, by our guard Carson who was very happy to be unexpectedly joining in the celebrations. It sure beats sitting around at the farm which is where he was meant to be!!!!

The bride and groom with family at the highland wedding with Dr Fikru. Also an Aussie back left!!!

Fikru proceeded to push people into groups for photos and all the rest of the community came to look.

An hour or so out in the lovely sunshine in a beautiful highland village in remote Ethiopia surrounded by scores of gorgeous people treating us just like family. It does not come any better than that.

We went back into the house, this time for lunch. Again, all seated on the seats and to our surprise the jug of water to wash our hands, the plates, injera, then wots were all brought around by some of the menfolk. This is the first time we had ever seen anything to do with meals helped out by men, but according to Fikru there is much more equality in the more remote areas. This meal was a lovely vegetable wot, and a separate beetroot wot that had no chilli. We had taken a large bottle of Tej as a gift for the family and I had a glass of that so that I did not have to drink the powerful araki spirit. We also had plenty of our own water, so we did not have to be concerned about drinking the local water.

After the lunch was cleared away out came a 20 litre plastic water drum which Pip held secure on the knee of one of the men and the African drumming began. Soon there were men and women doing traditional dances and lots of laughter and chanting. I was encouraged to join the dancing and of course my efforts of local dance, which involves rapid shaking of the shoulders combined with little jumps, had all the adults and children shrieking with delight and laughter. All the main regions of Ethiopia have their own dance style, and I tend to get them mixed up, so I suspect I was dancing more Tigray or Afar than Oromo. The truth is I suspect that I looked more like a copulating flamingo.

At about 2.30, Fikru had organized a van from Addis Ababa to come up the road as far as possible - and would we take a load of his family in our 4wd ute until we met the van. Most of the next hour we were saying goodbyes and listening to some of the children who by now had taken on the drumming of the plastic water drum, singing and dancing.

Eventually I coaxed the ute full of people along the rough track to the van where we had lots more farewells and we headed back to our farm with Carson - the farm guard who by now was pretty boisterous after quite a few araki's in the highlands!!

Chapter 11

Dec 2014 - Hardship and History

EE.39: A fine line between life and death

Pip and I were constantly amazed at how happy Ethiopians are in their day to day lives. Hardships are encountered regularly but the challenges are overcome, and it is always worth celebrating that you have lived to see another day. Unfortunately, disease, injury or death are also never far away which means life expectancy is about 58 years. This is skewed a bit by the fact that one in seven babies do not live past their fifth birthday.

Day 306

Increasingly frustrated at the lack of tap water yet again and we cannot find out what the trouble is. We do keep a good supply on hand for drinking, and we can filter more if necessary.

Tibebu with Getun driving, went in the ute down to Ambo to pick up the new overseer who is filling the position that Lechisa had previously held before he was poached to manage a section of a much larger property. They arrived back with Tibebu shaken to his core and announcing that he will never again get into the ute with Getun driving. It looks like Pip still has a bit of work to do there!!

The new overseer, Moti, seems like a nice fellow in his thirties with some experience in Natural Resource Management and a degree in Agricultural Economics. I don't think he has any agricultural experience so we should be quickly able to fill his mind with information.

In the evening, I made some stir fry which we shared with Moti and bread for Tibebu (fasting for him on Wednesdays) and we had a good chat. It was not long before Tibebu was at his raconteur best and had us in stiches with his stories. At one stage someone came to the gate and Tibebu went out into the dark and I could hear serious discussion. He came back to tell us that our dear old friend, the milking lady Gudatoo had just died. Different members of staff appeared out of the darkness for the rest of the evening all draped with their white gabis and slowly caught up on the news and the details. Tibebu said family members wanted Pip or me to drive to the Ambo hospital in the ute at 10.00pm and pick up the body. He reluctantly turned them down as he knew we would not drive in the dark. He advised them to get a contract vehicle from Ambo, and we would take up a collection to pay for it. Pip and I gave Tibebu 200 birr (A$10) for the kitty which was probably all they needed.

Apparently Gudatoo had bad gastric for a few days, and her family had taken her to the private medical clinic in Guder, but they could not afford to keep her there, so took her to the Ambo hospital. She was dead in 4 days. The consequences were sadly as Pip and I would have expected at the Ambo hospital. Death is so much part of life here.

EE.40: Friendship/Hardship and the DERG

Ethiopia has an enormously long history by our Australian standards with mythical stories dating back to the time when the Queen of Sheba sailed down the Nile for a quick "rendezvous" with King Soloman. So was born the beginning of the line of Emperors who ruled over the Kingdom of Axum, later to be trimmed down to the current Ethiopia. But there is also so much "colourful" recent history which is well and truly in the living memories of many of these people.

Day 329

All and sundry at the farm gave us a tremendous welcome back from our holiday in the UK for Christmas. As we expected, so much of yesterday afternoon and today was spent greeting people with endless hugs, kisses, and handshakes. It is such a good feeling for both of us to get such a welcome from people whose names we virtually can never pronounce correctly and with whom we share no common language, let alone any understanding of their (often troubled) pasts. Anyone more than 25 years old has been through periods of extraordinary repression and revolution with the associated fear and massacres which we cannot begin to imagine. Yet we are greeted with warmth and smiles that you would think have been sculptured in heaven.

Lots of visits to our hut from the local children and Pip was busy teaching them how to do jigsaws we had brought back from the Op Shops in the UK.

Pip with some of her "tribe" on the floor of our hut.

In the afternoon we went for a walk down to the river area and most of the eucalypt plantation had now been cut down and sold and sent off to Addis Ababa as scaffolding. It will rapidly regrow and be ready for harvest again in four years. We did see the troupe of Colobus monkeys who now do not have much shelter but fortunately there is still scrub and trees in the steepest areas and a few large Fig and Podocarpus trees for their food source.

In the evening Tibebu brought us some tomatoes and eggs from his farm next door and then sat and chatted until almost 11.00pm. He told us a lot about the recent history of the country and how he well remembered the overthrow of Haile Selassie in 1975 by the communist, Cuban backed DERG regime. During the DERG period he was a teenager and just managed to keep alive by getting a certificate from the head of his local Kebele to say he was pretty dumb and only completed Grade 6 - when in fact he was smart at school and had completed grade 11. If the DERG knew you were smart there was a good chance you would be killed or put in jail. They called into houses any time of day or night and took people away. Nobody trusted anyone.

Tibebu was later involved in the revolution to get rid of the DERG in 1991 which was a period of incredible bloodshed. Lots of people from around this area (particularly the children) were taken by their parents into Addis Ababa during the DERG period so that they could get lost in the crowd and had a better chance of survival. Hundreds of thousands of the best and brightest in the country were massacred in what is known as the period of "Red Terror" leading up to the overthrow of the DERG.

Tibebu is a great raconteur, and we really got him going tonight. After about 3 hours both Pip and I were thinking how incredible it would have been to have recorded the whole evening. It turns out that Tibebu had spent the last couple of years during the DERG time as quite an active revolutionary. During the day he pretended to be a dumb simpleton working in his uncle's flour mill. Each morning his uncle would cover him in flour dust and when the DERG militia men came into the mill he would stoop over his machine and grunt if asked questions. They would check his papers that said he had done grade 6 and leave him alone. He would spend the nights with fellow revolutionaries using a pistol they had "acquired" to shoot the windows out of army and police vehicles and buildings. Who knows what else, but maybe eventually we will hear more stories.

EE.41: Dangers

We did find that our survival instincts became much sharper as our assignment progressed and sometimes that saved us from unfortunate consequences.

Day 330

Tibebu and I headed off up the road in the ute to a funeral at Mutulu and to discuss issues with the water supply committee. We parked near the sawmill and after the funeral, Getun joined us, and I suggested that he drive home for practice.

Unfortunately, when we arrived for the funeral, I had pulled up a little incline to park the ute perpendicular to the road. When Getun was ready to go, instead of starting the ute, he let it start rolling back out onto the road. Of course, the steering was still locked as he had not turned the key, so he immediately panicked and slammed his foot on the accelerator instead of the brake. This did nothing to stop our increasing backward speed as the ute was not yet started. We were rapidly heading backwards across the road, over an embankment, through a house and so on down into a river. As we started down the embankment (there is no hand brake in the ute), I lurched into Getun's crotch and slammed my hand on the foot brake. We did stop before we hit the house but now, I had a problem. I am laying across the front seat with my hand on the brake and my head buried in Getun's crotch. Not an outstanding position to try and coach Getun how to get out of this predicament! Eventually, with my free hand I pulled his foot off the still depressed accelerator and pushed it onto the brake. This gave me a chance to regain my composure, and I coaxed him through starting the vehicle, and slamming his foot onto the accelerator at the same time as letting out the clutch. This I did by holding his knees and pulling and pushing accordingly. With a spray of gravel, we lurched back up the embankment and so home to the farm like a normal day at the office!!!

Farm cashier Kidist and son Lami had been away from the farm for a couple of weeks and this morning they returned. I can say that Pip and I were both delighted to see our little three-year-old friend - and he us. Pip spent most the rest of the day including him in her jobs around the house and veggie garden and showing him the new (second hand) books we had brought back from the UK. Later in the day, I had to go down to the grain grinding area and we headed off, me six-foot plus, Lami two-foot minus, me ghostly white, he as black as the ace of spades, his little black hand in my big white hand. True friends.

EE.42: Religious ceremony, Safety and Security

Day 240

Another extraordinary day in an extraordinary country. We understood it was a workday, but soon after eight in the morning, it was obvious that no one was coming to work, so it must be another holiday for the religious festival of Epiphany.

When Kidist came up from her hut to our courtyard, she said that there was to be a children's religious celebration with singing and horses and dancing up at the next village of Mutulu later today. We were welcome to come if we liked and we could all go together in the ute at 11.30am. It sounded like a lovely activity for the morning. One thing led to another and here is the report that Pip and I thought we should submit to AVI as we are under pressure from them to report any security incident no matter how small. It makes us a bit nervous that it could jeopardise our placement but feel that we should report so that lessons can be learnt by whomever needs to know these things.

Report of an incident in Mutulu:

This morning, we were told that the Epiphany celebration would be taking place in Mutulu, the local village about seven kilometres up the road towards the highlands. We were told to expect music, colourful horses, and a good deal of pomp and ceremony. As I was limited by an injured knee, we decided to drive and offered a lift to our farm cashier, Kidist, her three-year-old son, and a couple of older children. We also picked up a few other adults and children along the way.

As we drove, we passed many children and a few adults, all neatly dressed and walking towards Mutulu. About two kilometres before the village, we parked the vehicle and walked down towards the river, where the ceremony was already underway.

There were drums, singing, and chanting. Priests and religious figures were dressed in ceremonial robes, and groups of boys stood in circles, each holding a solid stick about one and a half metres long. They were chanting and mock fighting as part of the celebrations. There were several hundred people present, mostly children under fifteen, with some older teenagers and adults. To our surprise, there were also four local police and six federal army police present, most of them carrying machine guns. We commented that this seemed rather excessive for a religious event attended mostly by children.

After about an hour, the priests began their procession back from the river, returning the symbolic Ark of the Covenant to the local Orthodox church, as was happening in many places across Ethiopia that day.

We then drove into Mutulu and parked about eighty metres from the church. By this time, the crowd had grown to more than a thousand people, many of them youths. Groups of boys were moving around chanting, carrying their sticks, sometimes jogging up and down the road together while chanting.

Shortly after we got out of the vehicle, the crowd suddenly became much denser. Without any obvious warning, the police began using their batons to disperse people, and the crowd scattered in all directions. We did not see what caused this, and the disturbance ended as quickly as it began. At that point, the main procession was still about a kilometre away.

When the procession arrived at the church, Pip and I followed the crowd in, together with Kidist, her husband, and several boys from our farm staff families. After the ceremony in the churchyard, we decided to leave.

Earlier, the police and soldiers had been spread through the crowd in pairs. Now they had all gathered together in one place. This made me uneasy, and I said to Pip that we should leave. We walked back to the ute.

A few moments later, there was another disturbance in the churchyard. Hundreds of children came running down the road and across nearby open ground, shouting and yelling. We then heard a gunshot, which we assumed was fired by the police as a warning.

We had a bunna (local coffee) in a tent near the ute, which had been set up beside a small roadside stall. The situation seemed calmer. Kidist and her friend said they were going to get a drink and asked us to look after her son, Lami. We also had Barjar with us, the six-year-old son of one of the farm guards.

We stayed near the ute, surrounded by many children, most of whom simply watched us, while a few tried to talk with us in English. Some older boys we knew told Pip that there were decorated horses further up the road. She went with them to have a look and to find Kidist so we could head home.

I stayed at the ute, sitting on the tailgate, with Lami and Barjar in the back, while a small crowd gathered around. Suddenly, further up the road, fighting broke out again between groups of young men with sticks and the police and soldiers, who were also using their batons.

It was too late to leave. I crouched behind the ute, and about fifteen children jumped into the back and crouched there as well. About ten metres away, the police were beating one man with their batons when another group arrived and began attacking the police. The fighting then moved past us and further up the road, and more shots were fired.

I took Lami and Barjar out of the ute and moved into the coffee tent, ready to go into the shop behind it if the fighting came closer. People were running in all directions. After the shots were fired, the fighting either stopped or moved further away.

About half an hour later, Pip returned, and we decided to leave immediately and drive home.

We did not write this out of fear, but as a reminder of how easily situations can become unsafe in Ethiopia when large crowds form quickly. With limited resources for crowd control, the police often resort to force when events begin to get out of hand.

It did occur to us the invidious position that the police are in when these huge volatile crowds gather full of young people with little to lose and a desire to make something happen. The police know they can get no backup in less than a few hours and only then if their mobile phones are working. They have no equipment other than batons, stick and guns – no sprays, handcuff, teargas, radios. They have arrived by public bus as they have no other transport, so no divvy van to lock away troublemakers and no way of escape other than to their police station made of mud. We decided that in some circumstances turning to shoot at the crowd rather than above it would be a fairly easy decision to make!!

The violence definitely had nothing to do with our presence, but we often think about what impact our presence has from a social/developmental aspect. Certainly, we are the only whites/foreigners amongst the crowd of thousands. Almost certainly there has never before been a white person at the celebration in this community. We can guarantee that virtually every person at the celebrations would have noted our presence.

We spend the whole time engaging with people, shaking hundreds of hands, acknowledging elders, talking to children in English and giving them a chance to engage with a white person. We are offered sweets by many of the children that we know well, and kindness at all times - including being "shouted" coffee by the owner of the coffee tent.

Do they ultimately "need" this intrusion or, with the shrinking world we all know that we live in, will it happen regardless, and should we be at the lead, hopefully demonstrating that we foreigners can be people just like them - kind, compassionate, and open - just with a different skin colour?

Chapter 12

Jan 2015 - Farming Ethiopian Style

EE.43: Community Life

Day 342

I was out early as the Weekly Times rural newspaper in Australia had contacted me to see about a story and I wanted to Skype them before the end of the working day in Australia. At 7.30am the internet was hopeless, but I suggested to the journalist that if we waited until after 8am it might improve. Sure enough, it did improve and was then good enough to Skype. We did quite a long interview with a number of interruptions due to internet disconnections.

After breakfast I went off down the paddock and then I heard lots of children yelling in an unusual way.

I knew instantly that something was amiss and headed back to the hut. I found Pip rapidly gathering some things up so she could drive to Ambo in the ute.

It turns out that some young children shepherding their family cows in a nearby grass area had found a local boy of about fourteen hanging himself on a tree in a nearby paddock. The young children had cut him down out of the tree and started yelling across the paddocks. Adults rushed over and he was carried up to the road and by then Pip had the ute out and Tibebu and Kidist in with her as well.

They put the boy in the ute along with as many family members as could fit in.

He appeared unconscious but Pip saw his eyelids move so was pretty sure he was not dead. She tried to get them to let him stretch out, but they all squashed up close as usual.

So off they went straight to the Ambo hospital where apparently he had a tube inserted in his throat and was given oxygen. Pip did not have the customary wait at the hospital as Tibebu came out within about ten minutes and they went to do the shopping and other jobs.

When they were about to come home, they got the message that the boy could go home so they picked he and his family up again. Pip said he vomited into a bag on the way home. We later heard he had beaten up his sister and had tried to hang himself in fear of his father's retribution. Nasty business!!! Never a dull day.

EE.44: Land Title

In Ethiopia all the land is owned by the Government. Most is occupied by farmers who have been allocated areas, but some is common grazing land. Sadly, when the Government wants to do someone a favour, they hack off yet more of the common grazing land and give it to their mates. Also, during the period of the DERG, much of the land was reallocated and farmers who had developed their plots suddenly had that land taken away and were allocated another area. As most of the current local farmers were around this area at that time, they are still reluctant to undertake any infrastructure development such as irrigation because they worry that the Government may do another redistribution.

Day 347

We tie up Mr Ram on a patch of grass each day and then can leave Mrs Sheep untied and she stays around with her lambs. The labourers were busy preparing the threshing pads for the teff. They cleaned up the main threshing pad which has the 17 haystacks of teff around it, and then the girls mixed up cow manure and water in a wheelbarrow and plastered over the whole area which apparently seals it nicely and makes it easy to sweep up the teff grain eventually.

The men also chipped away the grass from the area which will be the maize threshing area, and the girls "plastered" that as well.

Threshing the teff using oxen

In the middle of the afternoon, I took the tractor and trailer down to the biogas area and started shovelling a load of compost into the trailer.

I thought it would be interesting to see if I got help from anyone but apart from a good help from Pip no one else showed up. When I got back to the courtyard Moti and Kidist were standing outside Tibebu's room and he was laying on his bed. Since I had seen none of the other managers all day whilst I had been busting my gut shovelling, I had a bit of a head of steam, I called across to Moti "what are your wonderful management team up to today Moti". The "wonderful" bit must have given him and Tibebu the idea that I was being sarcastic which of course I was.

Tibebu came immediately out of his room and started to tell me all the meetings that he had been having all day and how all the managers were away today and how he was about to go to his office and write down how it was part of their annual leave, and they were not just having a day off etc. As always, I did not pursue it and acted quite relaxed, and then he started to tell Pip and me what today was actually about.

We spent the next 2 hours fascinated, enthralled, and highly entertained. Turns out that the Ethiopian Government has finally decided to try and survey all the farms (there must be literally millions of them) and plots of land in the country and make a definite boundary.

The Government will continue to own all the land, but the current occupants will be recorded for each plot. It was time for our area to be GPS's today by four Government officials. It was done many years ago by pulling a rope around and people were issued with little books that gave vague ideas of boundaries, but according to Tibebu they were completely inaccurate and only vaguely resembled the reality.

This time, adjoining landholders had to agree on what they thought was the boundary and then the GPS coordinates were taken - and that now becomes the official boundary. Needless to say, as all of our managers have farms of their own, they were very keen to be at their farms today to put their case as to where their boundaries should be.

Tibebu had us in stitches telling us about some of the battles he had with different neighbours around our farm here during the day. In some situations, more and more family members would come - thinking that extra numbers would sway their boundary argument. Earlier in the day all the landowners had been told to go to the Kebele building and swear on the Bible (or, if not religious, on a special cultural spear) that they would tell the truth about their knowledge of the boundary. According to Tibebu this is about as good as telling them they should not steal from others. A completely novel concept!!!

At one stage Tibebu was up against about one hundred members of a family who were all talking at once and demanding ever expanding boundaries to their family land. He said that he managed to shut them up for a minute and told them (falsely of course) that the GPS machine the officials were using could also record the sounds of their voices and that he would take the recording to the court, and if any untruths were heard the person telling them would go to gaol for five years. Most the family members slinked off, and he had a perfectly amicable agreement with the family elder.

Tibebu has no faith in the accuracy of the GPS machines and says that it will all lead to more confusion. He said (and then much to our amusement and delight began to act out) how the official walked around one plot of our farmland and declared the boundaries were now official and gave Tibebu the area. Tibebu then asked him to walk around the plot again and put the GPS machine into his other hand on the outside as he walked around. Much to his delight (and our immense amusement as this was all recreated in front of us) he gained and extra 500 square meters for the farm. – There you go – vindicated his mistrust of the machine!!!

EE.45: Life on the farm – and still more frustrations

Management has always been instinctive for Pip and me since we started making our own farming decisions from about 20 years of age and we have lived and died by those decisions. Now I realise just how important management really is – the observations, the processing of the facts, the thinking about solutions and outcomes, planning, checking, double checking, and being really dirty on oneself for bad decisions, etc. How much of that could we instil in the people we were working with, whose biggest decisions in recent years appeared to be working out different ways to ask head office for more money so they can pay themselves their salaries.

Day 345

I was having breakfast when overseer Moti appeared and said that he had a call from Tibebu and could we go and pick him up in Guder. I drove down to Guder when I finished breakfast and brought Tibebu back to the farm. He was very agitated and fidgety, and I suspect was dying to tell me something about his visit to head office in Addis Ababa, and perhaps he wanted to give me a blast. I would not be surprised, as only a week ago Pip and I were with the senior managers in AGOHELMA head office telling them that the farm would never get going properly if Tibebu did not quit or pull his finger out. I suspect the managers pretty much handed on that message to Tibebu yesterday, so I was not surprised by his attitude.

When the power finally came on, I told Moti that the irrigation pump was fixed and was ready to go, and he agreed that some of the cabbages and chillis were a week overdue for watering but, ahh – he did not have any staff free. Maybe it was my head cold, or maybe just Ethiopian frustration (or maybe I am getting old and grumpy) but I was near Tibebu's office, and had forgotten about the mood that he appeared to be in earlier, as I stuck my head in his door and said, "so today we continue to fiddle whilst Rome burns!". "What?" was his response. "Well, you have six managers walking back and forwards between these offices, carrying pieces of paper and no one to do the watering". Ummm, it was not my best timing ever.

I could see him wondering if he should yell at me, throttle me, sack me, or just jump out the window. He turned his back, and I cleared out. Moti, who must have heard the "discussion," immediately found a guard and the watering began and by the time the power went off again tonight, the vegetables were all watered and looking happy.

There is an area of tomatoes that I suspect will get no more attention and after another picking or two, that will be it. That too makes me cross, as with a bit of fertilizer, and a continued watering and spraying program, some areas could go on producing saleable tomatoes for another month yet, as the plants still look very healthy and productive.

Towards evening I found Moti sitting quietly and just to make sure he did not think "life was meant to be easy" I asked him how his planning was going for getting the sweet corn planted. I knew dammed well he would not have given it a thought. We had all agreed in Tibebu's office at our planning meeting last week that some corn should be sown in the last week of February. He too looked decidedly like killing me would be a perfectly good option instead of planting corn - but he did say he would get onto it. How many more times will it need to be pursued???

Chapter 13

March 2015 - Governance and Relationships

EE.46: Embassy – worldly affairs

Never in our wildest dreams did we think that an agricultural mentor assignment in Ethiopia would find us sitting at the dinner table with ambassadors, consul generals and defence attachés but such was our opportunity. In fact, the diplomatic staff are very isolated in the Embassy and rarely mix with the local Ethiopia people other than local embassy staff, so we found that we had a lot to offer them in their understanding of how Ethiopia works, particularly in rural areas.

Day 385

Up soon after 6.00am to get our morning chores done and load up to head off to Addis. As we lock up our two rooms, we give them a good spray with insecticide to kill off any wildlife residing in the cracks and crevices.

We drove straight to the head office in Addis and picked up our fellow AVI volunteer, Gayle, and then drove on to the Hilton hotel. Now that we have our own vehicle, I am finding the driving in Addis okay - with Pip doing an excellent navigating job. If I just concentrate of avoiding everything and everyone around us doing ridiculously erratic movements in their vehicles or on foot, and get told "second exit around on the next roundabout", I seem to be able to keep up us out of trouble. Pip and I are starting to remember all the roads we need to get our way safely around the city. There are no signposts in the city and no roadmaps, so it is just a matter of "knowing" where you are going. We were a little worried all the way in today as our fuel gauge was getting down towards the E for empty pretty quickly and we asked at every fuel station but all the same response "yellum yellum" (not available). Eventually, on the edge of Addis, we found a fuel station with petrol and electricity to pump it and went through the torturous business of getting a turn at the pump.

At the Hilton Hotel we had been invited to a special event put on by the Australian Embassy to mark World Women's Day. It was for important people (like us!!) to have an input into directing the African Union on how to put plans into actions on the ground for the empowerment of women and girls. Australian Ambassador Mark gave the opening address, and the keynote speech was a lady from the African Union. Comments from the floor were then invited (Pip made the first comment of course) and then we had a very fancy lunch whilst continuing the dialogue at our tables. After lunch, the tables then reported on their discussion. They were the most interesting people to talk to for a few hours; a smattering of males but mostly they were the really important female movers and shakers in Ethiopia and beyond. On my table were the female Ambassadors of Finland, Ghana and Niger, the president of the Ethiopian Chamber of Commerce and two other important Ethiopian business women. All had reasonable English, so we had very interesting discussions. Pip had an equally interesting table with a man who is a program leader of the NGO Plan International, The Ambassador of Botswana, a senior lady from the UN, and a lady from the Ethiopian Education Ministry. We proudly sat there with our gold embossed name place tags saying, 'Agricultural Mentors.' Apparently, the event got good coverage on national TV the next night and Gayle and Pip both featured in full screen shots during the news item.

The event finished at about 3.00pm so we farewelled Gayle into a taxi to go back to work and rang our AVI manager Demiss We drove to meet him for coffee and a chat above a large supermarket where we then did most of our monthly shopping. We then drove up to the Ambassador's Residence where we had been invited to stay. The Residence is a lovely house and garden looking out over Addis Ababa and we received a very warm welcome and were put into a comfortable guest room.

We have got to know the Consul at the Embassy, Stuart Shaw, and he and his wife Linda joined us for dinner. Also present was Colonel Wes – the senior defence attaché for East Africa whom we have met several times. We had drinks with Mark and his young French wife, Anaise, then sat at the huge dining table with all the gold embossed coat of arms crockery and had a lovely dinner cooked by the resident chef. The nanny was fussing about looking after Mark and Anaise's three-year-old boy and three-month-old girl. Lots of lively chatter and funny stories and some interesting in-depth discussion about Australian foreign policy and where we fit into the world. A lot of very interesting views were expressed about China and the role it is playing and will play in the future of Africa.

EE.47: Ethiopian Democracy

Since the fall of the DERG regime and democracy being implemented, there had been elections every five years in Ethiopia. It seems that the same party always wins but we stayed out of arguments about the legitimacy of the Government.

Day 399

Back at the farm for a couple of weeks and this morning Pip was up early and headed out for some bird watching in the fresh damp air.

It was too damp early for teff threshing so the men repaired the bottom gate into the farm, and others were busy in the store with storeman Abissa making sure there was exactly 100Kg (one quintal) of teff in each bag and then sowing and stacking the bags. I got Moti to get a couple of men to broadcast (by hand) the oats and vetch mix onto the hectare that I disced up yesterday, and then I disced them into the soil. Also Pip and I and Abebech moved the chicken caravan again. We try to do this every month or so, which helps to minimize the risk of disease and gives the chooks fresh pasture to graze.

Pip and Tibebu keeping an eye on the chickens and the chicken caravan we built with proceeds from a grant from the Australian Government

Today, as some of the day labouring girls arrived and started packing corn cobs into bags, the police unexpectedly arrived and took two of them away on the grounds of suspicion that they had cut the water supply pipe a couple of days ago. They own the huts closest to the area where we have the most problems with the water supply pipe being damaged. We were totally shocked, and Tibebu is saying that they were only "overheard" to say it would be a good idea - but who knows. We presume they were taken to another big meeting at the Kebele today to try and sort out the water problems. Whether they were exonerated or further incriminated, we have not been able to ascertain.

Tibebu said that at the Kebele meeting, there was lots of agreement about getting the system working properly again so maybe all is not lost as far as tap water goes.

We did get a shower in our shower hut tonight, so there is some water coming down the pipeline. It felt quite lonely without an audience which we usually have when we shower under the irrigation bore!!. Pip laughed at my concept of no audience as she always has half a dozen sets of eyes trailing her back from the shower to our hut – but at least she had actually washed in private tonight for a change. Again, only about an hour of electricity for the day which at least gave us a chance to charge things up. There has been light rain most of the evening. The rain will be perfect for the oats and recently planted maize if it keeps coming. The men did get some teff threshed today as the afternoon was fine.

We had said that we will not go to Ambo tomorrow as we have been told to stay out of town. There is to be a huge meeting of politicians – something to do with twenty-five-year anniversary of multi-party elections in Ethiopia. The heads of all the political parties, including the Prime Minister, are coming to Ambo for a celebration. Security will be very tight, and no one really knows what will happen.

There is only one opposition member in the Parliament, so the concept of a “multi-party” democratic system is a bit hard for us to understand, but no doubt the Government has told the heads of the parties that they must be present or else they will no longer have “party” status. Then they would not have the right to stand candidates in the election.

Only twelve candidates can stand in each electorate, and if the “parties” put up a total of twelve candidates between them, then no independent can run. If the parties only put up say eight candidates, then the other four spots are drawn by lottery from amongst the independents that nominate.

EE.48: Local Women's Group

Certainly, one of our most inspirational days on assignment was spent in the company of many of the local women at a special Women's Group meeting day. Nothing could better highlight how effective self-help can be – nor how efficiently women in these rural areas can run an organization. It gave us considerable pride that many years ago, the Abebech Gobena organization had been instrumental in sowing the seeds for this wonderful group, and it has been carried forward ever since under its own banner. Yet another wonderful legacy from the lady Abebech Gobena herself.

Day 410

Just after lunch, Tibebu appeared and said that the women's group were waiting for us down at the Kebele. Now, what exactly this meant we had no idea as usual, but fortunately we got a bit dressed up and walked with Tibebu down the half kilometre or so to the school which also has the large Kebele "hall" which hosts all the community meetings.

It was the first time we had been in there and it is quite a large mud building with a corrugated iron roof. There were about one hundred and fifty women sitting on benches and a beautifully dressed woman giving a speech. There were about six men in attendance.

Of course, Tibebu, Pip and I were ushered right down to the front row seated onto a few plastic chairs. Moti had followed us down to the Kebele and came in a few minutes later.

The local women's group was started fifteen years ago by our organization getting some local women together and setting up a small financial fund. It has been supported along the way by Tibebu. Recently, it had won a large gold cup for being the best and most successful women's group in the Oromia region. There are about six main regiones in the country, so quite an achievement.

The group focussed heavily on developing a culture of financial saving amongst the women in the area. As virtually no local people have a formal bank account, the group acts as a bank, taking deposits from local women. Initially the women were encouraged to deposit half a birr a month but that has slowly been increased to ten birr (50 cents) each month. The volunteer group maintain a passbook for each of the women as most cannot read and write. The women can then draw out small amounts, or the group offers loans for women who want to start businesses. They started with 10,000 birr (A$500) from our organization and now manage 4.5 million birr (A$225,000) as they have built up their own group and also formed a cluster with three other groups from adjacent villages. It is all done locally by hand, by the volunteer committee, but the money is then taken to the Government bank in Guder who do the holding. Under the arrangements with the Government, the women are paid 7% on their deposits but can borrow the money for a business at 1% interest.

The group also started a heifer scheme similar to that we mentioned with the Mutulu group, where they were given (or bought – not sure) twenty-six heifers which were distributed to some women, and they have all calved and the calves are given to other women in the group. The group also helps households to try and get onto the water schemes and electricity and puts pressure on the Government bureaucracy for other infrastructure like a road bridge to the nearby Dogoma village. They are also heavily involved with the local preventative health and family planning service. One lady from the group (who used to work here at the farm when we first arrived and now works part time for the health service) also got an individual award for her efforts in maternal health in the area.

During the couple of hours, we were there listening to speeches all in Oromo, Tibebu's contribution and that of our organization were mentioned many times. He gave a short speech which he later told us was inspiring them to set out a further vision as they had been so successful fulfilling the vision they had held for the last 15 years. Unfortunately, there were a couple of men from the Woreda (next level of Government up from the very local Kebele) and they both spoke for far too long and according to Tibebu were talking "off the agenda for the day."

The six main women of the committee all sat out the front on the stage (which is just an extra level of dirt a couple of feet thick) which was beautifully decorated with some material, flags, table and chairs and some plastic flowers. Two other women were also up the stage making traditional coffee which was brought to us in the usual tiny cups at one stage in the meeting. As is customary, Tibebu was given his first being the most senior (age or importance) male present, then I was served and then Pip, followed by everyone else as the dozen or so cups were repeatedly rinsed in a bucket of cold water.

All the women at the meeting were beautifully dressed, most with white gabis over pretty, colourful frocks. It was no problem killing a couple of hours listening to speeches that we could not understand. We just sit there lapping up this incredible environment and admiring the six beautiful, serene women sitting on the stage.

I am sure all men and women from the Western world could just sit and admire the superb faces of the Ethiopian women with their large bright eyes, incredibly high cheek bones, perfect white teeth almost always with a smile and their languid, serene facial expressions. One of the great attributes of many Ethiopian people is their ability as orators and today was no exception. The men and women that gave speeches all spoke without notes and my old Rostrum critics training came back to me, and I could not fault the delivery of the speeches at all (apart from the men who went too long!!!).

At the local women's meeting

Just as the meeting was about to finish, Tibebu quickly explained that there was a tradition that visitors would make a small contribution, and with that he stood up in front of everyone and walked to the front table and handed the treasurer some money. The crowd clapped. Pip and I are getting savvy now with traditions and had both stuffed a few birr into our pockets before we left home just in case something popped up. So, I then stood up and walked out and handed in my contribution (more clapping) and Pip likewise. The meeting then seemed to break up as a few more of the visitors came up and made their contribution.

Tibebu, Pip, and I were led outside and offered the first take at the food – injera and a choice of three wots - spicy lentil, vegetable, and beetroot. Also, a bottle of soft drink. The other visitors and all the women then came out and joined in the meal.

Finally, we got back to our hut at about 4.00pm after another quite incredible Ethiopian episode. In the evening, we had a campfire and Tibebu and Moti sat and chatted with us until about 11.00pm. Tibebu told us that the lady who won the individual award has been instrumental in going door to door of every hut in the Kebele helping with women's health and family planning issues. Apparently, she had found "just" two women with fistula, and both had been sent to Addis for treatment and surgery and now both have jobs in Addis.

Dr Hamlin is an Australian woman who pioneered the fistula surgery and established the hospital in Addis. She is elderly but still active at the hospital which, over the years has treated tens of thousands of women and is still treating large numbers of women who suffer from this tragic ailment. For those who have not read the story, fistula often occurs in women who have a long and difficult labour and the pressure of the baby on the walls of the uterus causes a lack of blood flow to that area and if left long enough it can then rupture into the bowel or urinary tract causing lifelong leaking unless it is treated surgically.

Apparently, the stench is incredible, so most are left to "rot" in a hut for their entire lives. There are stories of women walking for days to get to the fistula hospital, but most are found by health workers and taken for treatment. With a lot of physio and with the surgery pioneered by Dr Hamlin it can be repaired sufficiently that the women can live a full "normal" life.

EE.49: Religion

We understand that around seventy percent of Ethiopians follow the Christian Orthodox faith, though their traditions differ noticeably from other branches such as the Greek Orthodox Church. In this region, a small number of people are Protestant, while across much of the eastern part of the country, Islam is the dominant religion.

Day 421 – Easter Saturday

Quite a few labourers turned up for work, so they were sent hoeing amongst the chilli plants. We are going to keep the chilli plants growing in the hope that we get more chillies from them in due course (which I doubt) and plant sweet corn in amongst them, so we have something coming on. The workers did half a day and then all went home to begin their Easter celebrations. We expected to see the animals being killed for tomorrow's feast, but nothing happened.

Late in the afternoon went for a walk down to the river area. It is so different now that almost all the eucalypts have been harvested, but it does give us a chance to explore areas that have been impenetrable in the past and we can get down to the river bank. Unfortunately, we still cannot get down to the bottom of the water falls and into the caves that we can see down there. It would be a great place to explore.

We could hear the Colobus monkeys quite close, and I guess they were upset as they have very little cover left.

The waterfalls at the back boundary of the farm. Note the cattle on the top getting a drink.

The men that have harvested the trees must be incredibly tough dudes. The path that they have used to carry out all the logs on their shoulders are so steep that we could barely scramble up them carrying just ourselves. Plenty of the poles they carried out were 12 – 15 meters long and 100 – 150 mm diameter. To look at the area through western eyes, you would say that it is surely impossible to get the logs out without any mechanical assistance. Not to mention chopping them all down and trimming them up with axes made from old vehicle springs.

In general, a large proportion of the community seem to be deeply committed to their religion. Apart from the Orthodox Christians thinking that the Protestants go a bit overboard with their yelling and chanting, Ethiopians in general are very accepting of the religion (or lack thereof) of others. This may well be due to their belief in the existence of the Ark of the Covenant, which they believe makes their religion more holy and closer to God than any other.

Replicas of the Ark are held in every church, so it plays a major part in their worship. No one other than the protecting Priest in the "Treasury" at Axum is allowed to see the original Ark. If they did see it, then we are assured that their eyes would burn up in their head!!!! We look forward to our trip to Axum in July to see the building where it resides: it could be the culmination - or the beginning - of our own religious pilgrimage.

For dinner we lit our campfire for some lamb chops, but did feel a little guilty having meat tonight when everyone else has been fasting for 55 days and they cannot eat meat until tomorrow, starting at 3.00 am in the morning.

EE.50: Local Hut – Relationship/choosing a spouse:

It was such good luck that our assignment meant that we were living in the centre of a fascinating community and that we had Tibebu frequently sit with us for hours around our campfire. We felt like we could ask him anything about traditional customs and life in general and we learnt so much during those discussions. Generally, Ethiopians appear to have no interest in anything outside of Ethiopia, but Tibebu was one who was keen to hear about our country and to discuss world affairs. He obviously watched enough television to know what was going on in the outside world.

Day 422 – Easter Sunday

We did not get out of bed until 9.00am. The temperature is so perfect every morning and the sun streams in the window onto our bed. So long as there is not someone outside the window trying to attract our attention for some demand or other, then it is a very relaxing spot.

Last year the slaughtering for the feast to break the 55 day pre-Easter fast seemed to be mostly done on Saturday ready to feast Sunday, but for reasons unknown to us, today seems to be the slaughter day so a lot of the staff turned up here to help slaughter a skinny old cow they had purchased from the market. Also, I noticed quite a few slaughtering operations going on as we drove up to Mutulu this morning. The fasting and then the slaughtering of animals and eating is an extraordinary cultural activity. Even Abebech, our chicken girl (who by Ethiopian standards is a little more sophisticated than most) told me today what a wonderful day it is as she had eaten chicken, sheep, and ox all in one day.

No water in the tap all day, so our first wash under the irrigation bore for quite a while. At 7.00pm we had been invited to the home of Aberash our farm shopkeeper and her son Hordofa who spends quite a bit of time here at our hut. We were served soup with large lumps of sheep in it and injera with cottage cheese. We have our own bottles of drink so that we do not have to drink the tella beer (in this case un-alcoholic as they are Protestants) which tastes like charcoal and is made with local water – two good reasons to avoid it. Under pressure to be culturally acceptable, we both did have a taste and confirmed our earlier beliefs!!!

Aberash has electricity in her hut and is very proud of her small, distorted, over-coloured TV with DVD attached (playing a Sherlock Holmes movie that nobody really watched). She also has an electric coffee grinder, a hot plate and an electric injera plate. Apart from these items of the new world, we pondered how to describe this hut. It is so typical of many others we have been in, and we do not want to sound derogatory – which is not at all how we feel when we are honoured to be invited for dinner, for bunna or whatever – as this is truly a home. The hut is made from cement and mud. There are no windows, just a door that is a couple of large slabs of timber nailed together and dragged open and closed with big sliding bolts of either side to lock during the night or when away from the hut. In the main area, the walls are mostly painted in blue but once through the curtain and into the other sleeping and cooking areas, the paint gets patchy and some of the walls are just raw mud.

The floor in this hut is mud underneath but has some broken vinyl on top. The lounge does not have anything other than the TV on a wooden table, a large eating table and about twelve steel-legged wooden chairs. In the corner are a few bags of teff and wheat stored ready for the rest of the year. Most the walls look like animals have been rubbing on them for years, and they need a good scrub.

Quite a few religious pictures and ornaments hang on the walls, as well as the very common black and white head portrait photos of men, usually the husband I presume - though they are so unreal it is hard to tell. We are not sure who they are in Aberash's case, as she has had at least a couple of husbands, but now lives alone. One of the former husbands came in tonight just as we were leaving. Hordofa introduced him as his father. You would need better than an excel spreadsheet to plot the genetic permutations and combinations of the children around this area!!!

When we have asked about the safeguards to in-breeding in the smaller more isolated villages, we have been told that you cannot marry anyone within seven family links to yourself. I have asked, how then do you find a wife if it is not organized for you by distant friends or relatives. The answer from Tibebu is that when you know a festival time (Meskel, Easter, or Christmas) is coming, and it is time for you to marry, you walk to a village maybe 100 km away and join in their festival and hopefully meet a girl you fancy and who fancies you. If not, you walk home and set off again at the next festival time. Sort of like dashing down to the nightclub (or swiping right or left on a dating app) but for these rural people, the commitment needs to be a bit more. A few days walk to an area you have probably never been to and then inviting yourself to their party!! Sounds quite fun though. Maybe it would seem a long walk home if you were jilted.

EE.51: The Hamlin Fistula Organization:

Yet another visit into Addis Ababa but on this occasion a stop at the Hamlin Fistula Nurse training facility. Then onto the first of many stays at the home of Stuart Shaw and his wife Linda.

Day 434

We expected a big day, and it has turned out as such. So many “experiences” in one day early on in our assignment and we would have been exhausted, but now we are so used to all that goes on just to stay happily alive for a day, we can take it in our stride.

We got away to Addis pretty much as planned at 8.00am with me driving our Landcruiser and Pip driving the farm ute with Abara the purchasing officer who was getting a lift to Head Office.

We called in to the new Guder hospital as we passed to see if the group of American doctors (whom we had previously met) were there for the last day of their weeklong clinic. They had just arrived in a minibus from Ambo, so we had fond goodbyes. Their organizing girl handed us a medium sized suitcase and a cardboard box full to the brim with all sorts of medical supplies. They had brought a huge amount of stuff with them from the US and were not going to use it all and she insisted that we would put it to good use. Numerous bandages, dressings, wipes, pain relief, eye drops, rubber gloves – even surgical masks which I sure hope I don't ever use!!! It is all invaluable to us as we still have medical issues to attend to in our community nearly every day and our supplies of antiseptic and dressings is getting low.

On the road in convoy towards Addis and I very nearly cleaned up two teenage boys who ran straight out from behind a bus onto the road. I managed to hit the brakes and horn at the same time, and I reckon they jumped so high they cleared the bus as they headed back to the side of the road. There seemed to be an inordinately large number of animals wandering on the road, vehicles charging up the wrong side of the road, drivers who clearly have not yet worked out which side they are even meant to be on, two year olds who find it more fun to play on the highway than look after their cows beside the highway which is what they are supposed to be doing.

We often saw trucks broken down bang in the middle of the road – not simply changing tyres, but frequently pulling out the back axle to change the back differential or removing springs which have collapsed, or other major mechanical repairs - with just a few strategically placed rocks on the road to warn approaching drivers that this obstacle is relatively permanent. We only saw one serious accident today.

On the edge of Addis, we stopped at the midwifery training centre for the Hamlin Fistula hospital. We have an AVI colleague Anne who has a placement there helping with training the lecturers and planning the curriculum. We had been meaning to visit for ages so today was a good chance, even though we had Abara with us and could only switch off the ute by parking it on a hill so we could roll start it again.

Anne gave us a great tour of the facility accompanied by a couple of the senior managers. They have a farm with about twenty dairy cows and a disused chook facility and a biogas plant. We had a good look around and could give them some good advice for their dairy and we discussed mobile chicken caravans which they thought sounded great to overcome most of the problems they have previously had trying to grow chooks for egg production.

Unfortunately, their poor cows had Foot and Mouth disease which was knocking their production around quite a lot. Can you believe it that the government vet had come some time ago for a FMD vaccination program and sure enough their herd, which had always been clean previously, suddenly got the disease. We can categorically say that the Government vaccination programs in livestock in Ethiopia are a complete disaster. It certainly makes you wonder about the human vaccination programs!!!

However, the hospital has a lovely campus which had previously been a military training area. Well maintained buildings, big old trees, ponds, dams, and green lawns. Ann lives in Addis and travels out each day but is very happy, so that was good to see. She does not mix much with the other AVI volunteers but is happy doing her own thing and has local friends and interests.

The campus also has some houses where most of the women spend up to two years after their fistula operations, getting their lives back in order and learning some skills to start life over again. Also, some of the women who cannot be successfully operated on, live on the campus, and have work there. We had lunch with Ann in the staff room and then hit the road again into Addis.

I transported one of the guys from the campus as he wanted to get into the city. He was the agricultural manager of the campus, so we had a good chat. At one stage he had tried to grow fish so, being a retired fish farmer, I could share a few ideas with him about why I thought it may have failed. The campus also had a bore with an inoperative pump, so again I could give them quite a few ideas. I think they will come out to our farm sometime soon for a visit.

We got into head office and dropped off the farm ute. The senior manager did not seem to have time to talk to us, so Pip and I hopped in our car and went around to fellow AVI volunteer Gayle's for a chat as she was ill and had taken the day off work. Her boyfriend Elias (Ethiopian) had told her he was going to work on Wednesday but had diverted to a huge demonstration in the middle of the city which was meant to be a commemorative display for the 30 Ethiopians who were killed by ISIS in Libya recently. Predictably it turned violent, and Elias has come back to Gayle very shaken up and affected by tear gas.

We then drove across town to the Kaleb Hotel where we parked whilst we went across the road and organized the purchase of another desktop computer and a printer for the farm – financed by friends in Australia. As it was getting late in the day, we then rang Stuart Shaw the Consul at the Australian Embassy. He had invited us to stay at his house, so we organized to drive around to the embassy and link up with him there.

We got there a bit after 5pm and had a chat with the Ambassador (Mark), and Stuart who told us how on Wednesday the Embassy had been very active getting an Aussie guy out of the clutches of the Ethiopian security forces. There is a young Australian couple who got a visa a few months ago to come into Ethiopia as journalists. He is a writer and she a photographer. They freelance, but file stories for masthead Australian newspapers and only got the visas on the grounds that they reported soft stories and did not go anywhere close to politics. We have met them a few times at functions, and it is likely they will come to the farm some time. Anyway, being a journo, the prospect of a good story is irresistible, so the young man snuck into the demonstration which Elias had attended. Inevitably he got too close to the action and got quite badly beaten up and put in jail for getting out his mobile phone and trying to film. Hence Stuart and the consular team had a busy Wednesday afternoon getting him freed. Time will tell if he and his partner are thrown out of the country, but it is quite likely as Ethiopia generally has a ban on outside media.

We left our car at the Embassy and went back with Stuart to his home. It is a lovely three storey home, very well fitted out with all the latest gadgets – even automatic taps in our ensuite. Nine guards rotate to keep an eye on it and the house next door which is also rented by DFAT. Your taxes at work!! Stuart's wife Linda is in Australia now visiting their daughters who are on school holidays from their boarding school, so Stuart seemed really pleased to have our company and we certainly felt very comfortable in his home. He cooked us a roast chicken for dinner, and we shared a very nice bottle of Hunter Valley red which had arrived in the "diplomatic bag." We checked emails for the first time for ages.

EE.52: ANZAC Day – exactly 100 years since the Gallipoli Landing:

Expats from all over the world try to maintain their links with home whilst living in Ethiopia. Diplomats are constantly being invited to share in the national commemorations of all the different nations and that is very much part of the ex-pat activities.

Day 435

We were up at 4.00 am as we had been invited to join Australians and New Zealanders for a dawn service at the Commonwealth War Graves Cemetery. We had a quick coffee and then went in Stuart's car at 4.30am, picking up another two Australians on the way. We arrived in the pitch dark and had a very memorable experience as described below. This is my special report on the ANZAC day dawn service – Addis Ababa.

The dark empty streets of Addis Ababa are a stark contrast to the usual hustle and bustle of humanity and vehicles during the daylight hours. We were in the car of the Deputy Head of Mission for the Australian Embassy, headed to the opposite side of the city to the Commonwealth War Graves Cemetery on a hill overlooking part of Addis Ababa at 4.30am on the 25th of April 2015- 100 years exactly after the Gallipoli landings. We needed to be in good time for the dawn service starting at 5.30.

To access the cemetery gates requires a drive up the wrong direction of a one-way road for a few hundred meters and we followed several sets of taillights up the road and into the carpark. The morning glow had not yet appeared on the horizon as we walked from the car park to the area for the ceremony with glow sticks lighting the path and dark shapes all heading in the same direction. My skin tingled with the chill in the air, the anticipation of what may be to come, and the slightly haunted feeling of seeing the dark shadows of people passing by hundreds of barely visible graves and headstones as we walked up the hill.

As we got towards the top of the hill some of the lights of Addis Ababa came into view, but it is not a city with much glow or twinkling of lights during the hours of darkness. The city sleeps, having somehow survived another day of hardships, deprivation, and chaos, interspersed with snippets of enthusiasm as new buildings are erected, more sections of the new railway system completed, and deals signed for industrial expansion.

A few muffled voices and the sounds of footsteps on the rock stairs shared the eerie silence with the distant amplified voices of the Imam with his "call to prayer" and the Orthodox priest delivering his morning words of wisdom as the communities waken from their slumbers.

We were welcomed into the area of the cenotaph by staff from the Australian Embassy and given a glow stick to hang around our necks. In total darkness except for these spots of yellowy, green light we huddled in groups with those that we had recognized, unaware of the lay of the land around us. As our eyes adjusted to the darkness, we could ascertain that we were to share the morning with about 100 others. The Deputy Ambassador recognized our small group of volunteers and welcomed us and said the program would begin as soon as he had located the Turkish Ambassador who was to deliver part of the upcoming program.

Bagpipes – there is nothing like the eerie moan of the first wind emitting from the bagpipes to waken the emotions, and in a few moments the two pipers were in full sound. Almost, as if on cue, what had seemed a pitch-black sky suddenly had enough light for us to make out the silhouette of the cross on top of the cenotaph. Our first sign that this was not a hill covered with graves but an area of dedication and grace, open lawns and manicured gardens, large old trees, gracious pencil pines, rows of white marble headstones and a centre piece of a large cenotaph topped with an imposing cross.

As the morning light unfolded across those of us on the hill and the city below and beyond, we heard fine words of prayer, of remembrance, of commemoration, of camaraderie, bravery, spirit, mateship, of nations fighting together for a common good and nations fighting against one another for no good. The beautifully spoken words of the Turkish Ambassador reflecting on the event of the Gallipoli landing as a tragedy for all, from individual families to whole nations, had my rather feeble emotional barricades soon broken and I confess to shedding a few tears onto the grass of a hill overlooking a city and a country a very long way from home. Homesickness or a sadness for lives lost - probably a bit of both. Hours before, as dawn broke over beautiful Australia and earlier New Zealand, tens of thousands of Australians and New Zealander's had no doubt also been reflecting on the happenings on the beaches of Turkey one hundred years before, and on so many other conflicts before and after that have helped shape our nations, but along the way have created much sadness for so many families.

The Australian defence attaché, Colonel Wes Volant, coordinated the half an hour or so of speeches and prayers and the playing of the national anthems of Ethiopia, Australia, and New Zealand. By now the morning light had joined us on the hill and the scene of military personnel from many countries in full dress uniform standing at attention and saluting as the anthems played was very spectacular. Admirals from the US Navy stood alongside brigadiers from Botswana, colonels from New Zealand an Air Vice Marshals from Turkey and other fine, highly decorated military people from nations near and far. Ambassadors, or their representatives, joined Australians and New Zealanders and their families who are stationed in Ethiopia - and a small contingent of volunteers dressed in our best (not very formal) attire as we all absorbed the emotion and sense of occasion.

The anthems were followed by the laying of about 30 wreathes, mostly of yellow or red roses in glorious arrangements which were placed around the cenotaph by senior representatives of the different countries who had gathered for the occasion. As the first of the sun's rays reached the cross on the top of the cenotaph a trumpeter played the last post and, following the minute's silence, the reveille. Simultaneously, three young children who are family members of some Australian and New Zealand embassy staff raised the flags of Ethiopia, Australia, and New Zealand on three small flag poles.

Proceedings were finalized by a prayer from the presiding minister of religion, and now in full daylight we had a chance to inspect the wreaths, gardens, and headstones in our surrounds. There is one Australian buried in the cemetery and a special wreath had been placed on his headstone. A young man, Flight Officer Bull who had been killed in an air crash over Ethiopia, the only Australian military person to have lost his life in Ethiopia during conflict.

By 7.00am most of the participants (about sixty Australian and New Zealanders) of the dawn service had found their way to the Hilton hotel where we were joined by another 100 or more who had not managed to get to the service, and there we had a huge "gunfire" breakfast. A very lavish breakfast complete with the traditional bacon and eggs, Anzac biscuits and coffee laced with a dash of Bundaberg rum. The New Zealand ambassador gave a short speech and some of the livewire young Australians and New Zealanders soon had the two-up pennies spinning with the appropriate raucous yelling and laughter.

Great fun and of course we volunteers were pretty much the last to leave!! We did not want to spoil our reputation of being the ones to suck the last bit of fun (and maybe food and drink) out of every occasion.

By now we had been in rural Ethiopia living in our little hut a very long way from anywhere for fourteen months, so we greatly appreciated the opportunity for living some of the high life. But always after a day or two we were keen to get back to our community. Our AVI manager Demiss said he really could not understand us. He would offer us some AVI sponsored R & R in a five-star hotel but after a day or two we would be on the phone to him suggesting it was time we got back to our hut on the farm!! We were on an agricultural mission, we loved it, and we were not going to be distracted!!

Day 436

After a good sleep, a hot shower and a late breakfast in this lovely home in Addis Ababa, we were into Stuart's car and headed to pick up Colonel Wes at his huge home on the hill almost next door to the Ambassador's residence. Wes is the senior defence attaché for a large part of Africa, and we get on well with him. A veteran of almost every conflict that Australia has been involved in recent decades he has a great sense of humour, is very generous and obviously highly respected - and a mad keen Richmond AFL club supporter. From his huge home we headed to the British Embassy to attend their annual gymkhana. The embassy is so colonial, a huge area of 40 hectares perched on the side of the hill overlooking Addis. The high-tech fence around the perimeter with guard posts every few meters is four and a half kilometres long. It is the biggest area of any British Embassy in the world and has dozens of houses and buildings. In the centre some beautifully maintained round tukul houses with thatched roofs which were the original embassy buildings when the area was given by Abyssinia to Britain in the early 1800's.

Some of the buildings at the British Embassy

The event had started earlier in the day, so we were not sure about all the activities, but there had been a dog show (expats just bringing their pets really) and face painting etc for kids as well as about fifteen young people from the pony club on the compound who competed in some gymkhana activities. I am sure there was at least half a dozen fancy ribbons awarded to every contestant. The Poms really do these things with style. We were just in time to get some of the last scones with jam and cream.

As the events wound up at about 3.00pm we had met the head of security for the British embassy, a retired English senior police officer. He invited Stuart and Pip and I for a quick tour of the grounds and then back to his residence, a glorious old colonial weatherboard house in the grounds of the embassy, surrounded by green lawns and huge old trees. Only the peacocks were missing from the scene. (There were about ten brown chooks instead).

We sat on the terrace drinking large amounts of gin and tonics and eating salted peanuts and crisps – such a treat for us - we had to apologize for our bad manners at consuming whatever was on the table. Alan's wife and daughter joined us, and they are generous, friendly people – obviously with fascinating stories to tell as to how they ended up in Ethiopia. A quick gin and tonic turned into about 3 hours of delightful chat and finally we headed back to Stuart's house, but only after Alan's wife had loaded Pip and I up with a kilogram block of tasty "proper" cheese, a block of "proper" bacon and other bits and pieces of food we never see. Yum – what good friends to make!! Rationing of these delicacies will be fully implemented when we return to our hut!!

Back at Stuart's I cooked a huge fillet of beef that he had taken out of the freezer. He had "picked it up at the airport" as he flew home from South Africa recently and has been looking for an opportunity to eat it - and we apparently were a good excuse. I roasted it with potatoes, onions, and beetroots and cooked cauliflower with white sauce. It was a wonderful ending to an incredible weekend.

EE.53: Restructuring the Farm

Not a day went by without Pip and me feeling frustrated as we watched guards and managers sitting idly, doing very little. Much of this seemed to stem from long-standing traditions—or perhaps even laws—that kept the system in place. For example, if someone was appointed as a driver or cashier, we were told they could not be asked to take on any other tasks. We were never entirely convinced this was truly law, but it was certainly treated as such. Likewise, the chain of command meant that once someone was appointed as a manager, they were not expected to actually manage. Responsibility was always pushed up to the next level, leaving no one clearly accountable. And to make matters worse, we were told the only two grounds for dismissal were theft and drunkenness—ineptitude and laziness could only be dealt with through endless "counselling."

After we restructured roles at the farm, though, things began to change. Head office was so impressed with the results that they started reviewing every department at head office and rewriting job descriptions. From then on, each employment contract carried the crucial phrase "or other duties as directed." It may sound like a small adjustment, but for us it felt significant. If we achieved nothing else during our assignment, this single change might stand as one of our most lasting contributions to the efficiency of the whole organisation.

Day 451

A day to really test our management skills. Soon after breakfast Tibebu arrived at the farm and asked that we have a meeting with Moti to resolve some long-term issues with the farm and then he would report to Head Office as to the future direction.

The four of us ended up meeting for three hours in the morning and another couple more hours in the afternoon.

Tibebu opened up with quite a long monologue of where things were at and what needed a decision. Fortunately the issues he raised were almost exactly along the lines of Pip and my recent report to head office. It made us realise that finally he has got the message from us of the types of things that HAVE to happen for the farm to be financially successful. He made it clear that there is no way the farm can continue to keep paying all the managers to do almost nothing.

We then worked our way through every member of the management team and redesigned their job descriptions. Kidist for example, who has only had a couple of hours of work to do each week as cashier – but is paid for the full forty hours, would now have extra duties and be manager of chickens, checking health, and making sure the feed is on hand. She will also do the morning work on the chickens which includes collecting the eggs. Later when the chooks are let out of the caravan, a day worker will take over, as then the chickens need to be watched to stop them straying too far and to watch for dog attacks. Kidist would also take on a new role as office manager, ensuring all the paper trails are complete and up to date as well as begin to get some experience writing the endless letters that must frequently be sent to head office.

The purchasing officer (who currently has even less to do than Kidist each week) will now be expected to actually do the purchasing, relieving Tibebu of chasing up everything. He will also have added to his duty a new desperately needed role of maintenance manager. He will be expected to put together an audit of the status of all the buildings and assets and plan their maintenance. Unfortunately, he is pretty useless with his hands but will be expected to get day labourers as required to help him.

And so we went on through all the staff, including a revamp of the guard's duties. Needless to say, most of the ideas had come from Pip and me, but we were delighted how quickly and happily they were embraced by Tibebu and Moti.

Eventually when it looked like we were nearly at and end, I "bit the bullet" and said – "and Tibebu, what about you and Moti?"

I nearly fell off the chair when he then announced that the farm could not afford them both. With admin staff expenses of close to 40,000 (A$2000) birr a month and only 200,000 birr (A$10,000) in the bank he said that the farm will likely run out of money by October. I didn't actually agree as there will be some cash crop income of cabbages, sweetcorn, and potatoes by then, but at last he is seeing the reality. He then said that we would need to fire four other managers just to make the saving of his wage, and as his leg is always causing pain and his wife is sick of him being away from their home in Ambo, he had decided that he could hand over to Moti. Wow, this was so much better than us having to pour out lots of blood to get him moved on. I suggested that for a year or two he goes to one or two days a week (which is all he has done for years anyway) and I think he will take up that suggestion and hopefully come on a regular day or two to the farm. That will save the farm maybe seventy thousand birr (A$3500) a year and we will probably get the same amount of work, but it will be more in a consultative role than decision making (or lack of) about everything.

Win, win, win, win - as we see it.

We then went through ways to increase the turnover (and hopefully profit) of the farm. Pip and I had presented the idea that the farm utilizes all our administrative staff by revamping the shop and setting up a brokerage business for agricultural products in the area. We envisage buying the individual bags of teff and other grains and maybe vegetables from small farmers who currently have to load them on a donkey and take them to Guder to be ripped off by merchants. We would then aggregate the produce and send it to Addis, either to be paid for and used by the head office income generation activities or sold there for a much higher price than our local price.

We would also begin to sell small lots of teff and other grains to locals around here directly from the shop - either ready ground into flour or as whole grain. Tibebu says that 80% of the food consumed in this area is actually brought into the area and only 20% grown here, so there should be a good retail market as we are seven kilometres from the nearest shop or market. Also, we suggested that the shop may begin to sell hardware supplies like nails, corrugated iron, cement etc which currently have to be bought in Ambo and carried by people or donkeys back to this area. All of this would greatly increase the work for the cashier, purchaser, shopkeeper, and storeman and maybe we would begin to get value for the money we spend on their wages.

Finally, we discussed the current enterprise mix and Tibebu and Moti agreed with our proposition that the irrigation water should be concentrated on an increased area of sugar cane and that dry season vegetables – whilst having the potential to make good profits – require a much higher level of management and much greater risk so they should not be included in the mix. Also, that some of the cropping area should be turned over to production of fodder for the dairy enterprise.

We believe that with reasonable management the dairy is the highest priority and can make seriously good money and that dairy produce will continue to increase in value in the country as incomes rise. Currently there is almost no milk or dairy product available in the Guder or Ambo area with well over fifty thousand residents, an enormous university, a large ammunition factory with thousands of workers and an army training base which sometimes has up to ten thousand recruits in residence.

Our priority for focus on the farm land (and irrigation water) thus becomes the dairy, sugar cane, wet season cabbages (because my instinct says that it is pretty risk free and should be quite profitable if sold properly), maize and then teff. The areas that are currently cropped to potatoes, niger seed and wheat would be used for fodder production for the cows, using oats and vetch for rotation purposes. If we needed more fodder, then the teff area would be reduced as it is not very profitable even in a good season.

The revamped shop, grain grinding service and niger seed oil extraction (from purchased seed) will be continued.

All of this must be submitted to the general managers in head office for their approval. If we can pull it off, I think we can be proud that ultimately we will get the place restructured and onto a decent financial footing. A lot will now depend on Moti, and we will be busy for the next eight months getting him into his new role – IF the restructure is approved.

Quite a bit of noise from the sheep houses tonight as Pip has weaned her two lambs today. She had them on ropes of their own today away from Mrs Sheep and tonight they are in a different house. Poor Mrs Sheep is taking her chance in the same house as the seriously oversexed Mr Ram for the night. We hope she can get some sleep!!!!

Our "grandson" Lami feeding Mrs Sheep

Chapter 14

May 2015 - Ethiopia's Future

EE.54: Racism, Hope, Opportunity:

We had not realised how easy it would be to maintain a strong moral compass despite living in a community almost on the other end of the financial spectrum from the Australian community in which we have lived our whole lives.

Day 454

Today Tibebu is supposed to have gone to the town of Ginchi which is about 40 Kms away to see if he can purchase good quality potatoes for planting. Every town seems to have one main market day and often a second mini market day. Thursday is the day for Ginchi, a town through which we regularly drive on our way to Addis. Monday is Guder's market day – which incorporates the big livestock market as well, and Saturday is the main market for Ambo, with Tuesday and Thursday the mini markets there.

As I sat with young Barjar on my knee yesterday on the veranda of our hut, I could not help but think how good it would be for the world if I could have with me anyone that harbours any thoughts of racism or elitism. Here is a dear, young, jet-black boy who is almost as close to the bottom rung of the ladder of humanity as anyone can be. As the seventh son of our guard Abara, he is given a really hard time by all of his brothers, so definitely the bottom of that family barrel. His father earns about $A1.50 per day working 24 hours on / 24 off as a guard here at the farm, so almost as lowly paid as you can get to sustain 9 (surviving) children. Our community, being in rural Ethiopia, is a long way from the top of Ethiopian society, and of course Ethiopia is ranked in the world somewhere at about 147 out of 150 nations for income and net worth of its residents. So dear little underfed Barjar has pretty much just "survived" his seven or so years. Always in rags, always hungry, probably rarely, if ever, had any affection and yet he sits on my knee with a beaming grin, a perennial runny nose, and a glint of mischief in his eyes befitting any boy of his age. He has now spent enough time in our hut and with us for us to know that given an opportunity he, like most other children that we have anything to do with here, could be as successful as any person on this earth. So, who has the right to think they are "better" than anyone else?

We are pleased to say that Barjar does (occasionally) go to school and Pip now has him writing the alphabet letters –which he did much better this morning after she promised a banana if he concentrated on his writing for a while (the scientific approach helps!!). Generally, if he is not under our feet, he is standing at our door just staring at our banana supply or pantry shelves. He gets on really well with Lami, so they are both here a lot. I can't help thinking how he would bring the house down as a Flying Fruit Fly circus performer, but sadly it will never happen.

I should say that his father is one the of the most honest, considerate, and helpful men on this earth and he has been a tower of support for Pip and me, although we share not a word of common language. How often does he "sense" that we would like to roast some maize, and suddenly there is a little campfire outside our hut, or he returns Mrs Sheep to us if she has headed off for an adventure. The sort of man you would truly trust with your life, and we reciprocated by doing whatever we can to help his family in various ways.

A regular evening for us means that at about 5.30 pm, I head out to the internet spot. As there are almost always frustrations to be overcome with poor connection, it is not a particularly good place for Pip. I get back about six and then I head off to the shower hut about two hundred meters away with a shirt on top and long skinny white legs sticking out under my towel on the bottom half. Ah - the refreshing cold water pours out of the shower head – if the water supply to the area is working. About this time Pip heads off to get her sheep from wherever she has tied them up to graze for the day. She usually brings them closer to home and ties them up again around our courtyard. When I get back, I pour a stiff whiskey and pull out the yummy, freshly roasted peanuts that we buy in Ambo. I read the emails and any news articles that I have downloaded from ABC "Just In" website. Meanwhile Pip waters the veggie garden or goes bird watching until it is nearly dark at 7.00pm when she puts her sheep away and then heads off for a shower in her silk bathrobe and thongs. When she returns, I quit the computer, and she reads the downloads and emails whilst I cook the dinner on the veranda. After dinner Pip does the dishes and I return to the computer where I pass the evening whilst Pip reads books, plays Candy Crush on the iPad, or continues with her cross-stitch tapestry. Sometimes we watch a movie or doco from the external hard drive we brought from Australia loaded with movies and music.

EE.55: Disabilities and Hardship:

There are many levels of hardship in this world and sometimes when we are back on the farm in Australia and a drought is in full swing, the crops are failing, a tree has fallen on the fence, the old ewe is stuck in the cattle grid, and we have been working in the sun all day, we think we are doing it pretty hard. But really – where does that fit on the scale of real hardship?

Day 471

We are in Addis again as our AVI manager Demiss says we must not be at the farm during the Ethiopian federal elections.

After breakfast Pip and I needed a good walk so we rang defence attaché Colonel Wes to see if he was home and would welcome us for a coffee. It suited him fine, so we walked to his huge house.

We had to walk for half an hour or so through one of the most challenging areas of Addis. Not so much from a security point of view (although the unwary almost certainly get robbed) but it is just the rawness of Africa right in your face.

Around the Megenagna market area and bus stop, seething with people, hundreds of buses with "conductors" shouting for business, people and buses negotiating an impossible route to freedom, and everywhere the most desperate of beggars with every possible disability imaginable. Men lie on the ground in rags and hold one leg up in the air so that you almost walk into it, and instead of a foot on the end of the leg there is a gangrenous lump or simply a red scar where presumably a foot once was attached. Women with babies, skeletons with HIV, polio affected children sitting in or beside all forms of wheelchairs – and so much more.

As we started to climb the hill to the mansions on the other side of the main road, I said to Pip that the previous half an hour would be enough of the "African experience" for many people in their lifetime. Strangely, it felt to us just like another "day at the office" - but it is not another day at the office, and it never will be, or should be, to people from developed countries. Truthfully, it is a human tragedy on so many levels which will never be solved with overseas aid, but certainly the wonderful world-wide efforts to eradicate polio and control HIV/AIDS, and maybe eventually some workplace OH&S awareness in the country will improve the situation. It will be interesting to see whether we are scarred, or enriched, in the long term by experiences such as these.

Colonel Wes is such an interesting man, and we had an hour-long discussion over coffee, mostly on the dreadful situation in South Sudan (as well as the Richmond footy club and Australian cricket team). He had been in South Sudan during the last week with a small contingent from the Australian Embassy in Addis. With virtually no effective Government, the country is currently a complete disaster with atrocities to rival those of Rwanda decades ago.

EE.56: Marriage – and some Positive Farm Management

As we understand it from Tibebu there are three types of marriages in this part of rural Ethiopia – particularly within the Christian Orthodox community. The first is performed by a very senior church figure, supposedly between two "pure" people. It can never be annulled. The second type is very similar to our marriage but there does not seem to be anything more formal than a signing of a document agreed to between the two families saying what each party will bring into the marriage. The third type is described below and is usually for very young people – arranged or in love (or perhaps hopefully both!!).

Day 487

Lovely fine sunny morning again and plenty of action around the farm. Still teams of men and women hoeing maize and also some new workers beginning to fertilize and hoe the potatoes.

The oats and vetch were broadcast on their paddocks and disced in, and the new Kale seed was broadcast onto a bit more than half a hectare at the end of the potatoes. Tibebu was keen that we plant Kale as apparently the head office in Addis need a certain amount of Kale seed each year as they use the oil from the seed to "condition" the injera-making hot plates. Given that the injera-making business at head office make two and a half million injera a year, there is a fair need for the Kale!

Pip and I planted more beans, and I kept an eye on Getun who was driving the tractor. I had him pulling the smudger behind the tractor to flatten out the area for the Kale. It was the first time he has driven the tractor with something behind that he could not just reverse, and as I predicted to Pip, he eventually drove into a corner and could not get out. I had to go down and extract the tractor and smudger. That old 'thinking ahead' process we talk about so often does not seem to sink in - If I drive into a corner and cannot turn around and cannot reverse, then.............???

Late in the afternoon whilst we were planting beans, there was the usual thunder and lightning but then the heavens opened. We sat under a tree for a few minutes and then decided that it was going to be serious rain – and the start of the long rainy season. I ran right to our hut in case the rain went in under the doors, but Pip sheltered in a shed. She had fallen on the run back and hurt her knee. When she eventually got back to the hut, she put ice on the knee. Tonight, she says it is fine. My indestructible girl!!

We ended up with more than an inch of rain which will be good for most the crops.

Tibebu has taken most of his stuff from the farm to begin his retirement but came in for half an hour today looking for a personal bank book or something. I managed to have a chat with him, and he seemed relaxed. I said we were confused about the wedding/ engagement up the road planned for Sunday. He said it is the third type of marriage and the previous weddings that we have attended are the more normal second type. He says the girl on Sunday was only fifteen, but it is called a promissory engagement.

If the family (for an arranged marriage) or the young couple are determined to cement the union, then the engagement has all the promises that go with a marriage and will ultimately be fulfilled, but the girl continues to live at home. The couple are not expected to live together and start a family for quite a few years. This ensures (mostly) that the girl can finish her education. Tibebu does not approve as very often the couple start living together and, in his words, "do not keep the promise" – presumably of celibacy, and the girl gets pregnant and never gets educated.

Tibebu, as a respected senior of the community, advised the father of this girl not to let her get engaged or married until she has finished school or maybe university and he also gave the girl a good "talking too" as he has known the family for twenty-five years.

Young couple at the "promissory" wedding in a nearby hut.

It seems incredible that it is not much more than a couple of weeks that we all accepted the idea of Tibebu retiring, and now he is gone. Today Pip and I heard something that made us even more pleased with the change to Moti as farm manager. Earlier he was down in the farm shop when the rain set in and a few workers from the Holeta agricultural research station had stopped whilst driving down the road and gone into our shop to buy cheese, butter, and Niger seed cooking oil. Moti had started to talk to them and later reported to us all the information he has found out in just half an hour. What pleases me is that firstly he bothered to strike up the conversation with them but also that he then asked the right questions. Tonight, we are feeling very positive.

EE.57: International Aid

Of course, one would be naive to think that International Aid given by Governments is purely for humanitarian purposes. The main driver is so often politics and diplomacy but that is pretty hard to accept when you are living at the coalface and giving up a period of your life to help the developing world.

Day 487

Today we had a bit of an unsettling email from the Australian Volunteers International head office in Melbourne. Due to the Australian Federal Government (under Prime Minister Abbott) cutbacks to the aid budget, particularly to Africa, the Ethiopian AVI volunteer program is to finish next June. Fortunately, it will not affect our assignment, but it does make you feel as though the efforts of those of us in the country are not really appreciated. It just reinforces my theory that when the money is allocated in Canberra, it is entirely for political purposes but by the time we spend it on the ground, our intentions are purely humanitarian. As it filters down from Canberra to DFAT, to AVI Melbourne, to our main East Africa office in Tanzania and then into Ethiopia, the purpose of the money slowly changes. So, a minor political decision eventually has a significant humanitarian consequence. Have another wine in the Parliamentary dining room in Canberra I say and don't waste your time thinking about such things!!!!

Tibebu was in for a chat today which gave me a chance to run a few things past him, and I was pleased that he wanted to tell me about his own farming pursuits which was good to hear. He has a farm next door to our farm and is looking to build a house there so it is great that he will still be a part of the local community.

EE.58: Research and Science – and Local Justice

Still the mysteries continue even though we had now been on assignment nearly eighteen months. To discover that there was a significant agricultural research station close to the farm was definitely a major surprise, but a very, very welcome one as we pushed for better seed and animal genetics and lots of other possible productivity gains.

Day 525

Our plan for the day, as organized by Moti, was to go to the Ambo Agricultural Research Station and see is we could get a young bull to grow out for our future requirements at the farm. We have no idea how Moti knew about the Research Station as Tibebu had never mentioned it to us. We headed off about 9.00am with Moti and Getun with Pip driving down a very slippery road to Guder as we had heavy rain overnight.

We turned up the Wenchi Crater road in the middle of Ambo and bounced our way up the terrible broken-up bitumen in first gear in the Hilux – some seriously rough driving conditions. Bear in mind this is the road to one of the top seven tourist attractions in the country if you believe the tourism advertising. Ethiopia is "supposed" to be relying heavily on tourism, as the economic growth from building unoccupied concrete buildings in Addis is recognised for what it is – anything but true economic growth!!!

A few kilometres out of Ambo, we entered through the usual large steel gates (95% of all steel in Ethiopia is surely used in building impressive security gates!!) and the world was transformed. A lovely tree lined driveway and parklands largely planted with Australian natives – grevillea, silky oaks, casuarinas, eucalypts, white cedar and more that Pip could no doubt add.

We parked outside a large, multi-storied, quite tasteful concrete/ tiled building and were greeted by a staff member who took us up to the senior manager's reception office. He appeared soon after and took us into his office. A delightful, intelligent man with good English who gave us a particularly good welcome and was most interested in the fact that we were at the Abebech Gobena farm. His organization has worked with our farm some years ago and he has often bought our butter, cheese, and eggs. Why Tibebu has not ever mentioned this place to us is a complete mystery.

The national priority of this station is researching plant diseases and insects, but they also bulk up different crop seeds for the breeders at Holeta Research station. By the look of the offices there must be dozens, if not hundreds of scientists working here. According to the manager, they work closely with several American universities and have visits from American scientists almost every month. We went to their cafeteria and had milk coffee, and whilst we were there the manager explained that their dairy herd was very small, but they regularly sell their bull calves. All their milk is sold to staff.

We then saw a few calves about three months old and had the option of purchasing the males, but the manager and the dairy manager thought that we would be much better trying to get a bull calf from Holeta Research Station, as the stock at Ambo are inbred and only about 60% Friesian. At least we now know that the option is there if we cannot get a better calf somewhere else. The bull we have at the farm (purchased eventually by Tibebu from his own farm) is a good bull but the management now is a bit worried that in another year or so he will become too big and "aggressive". The best option is to start growing out a bull calf to replace him in a year or two. Some forward thinking in action....

We left the station after a couple of hours all really impressed by the way it appeared to be run and certainly impressed with the senior manager who told us he has been there fifteen years. He is a microbiologist and has a keen interest in the environment and conservation. He and Pip had a good conversation about the environmental problems that we had seen in the north of the country and also the problems developing locally.

Pip and I had taken a drive on the main highway about another 30kms past Guder and up into the highlands some months ago and were amazed and saddened to see even less trees and more erosion than even in this Birbirsa area. The manager told us that it is now like that for the whole five hundred kilometres to the Sudan border.

Fifteen years ago, when he came to the station and travelled west, there was a healthy natural environment almost all the way. That it has changed so much is a complete tragedy, and we cannot help thinking that far worse catastrophes are in store for Ethiopia than have ever occurred in the past.

Sometimes we think that the Ethiopian thinking goes like this: – Keep on making more farmland from the forests to produce more food and cut down every tree for firewood and charcoal, trying to keep the 90 million people alive and to save some of the 300,000 children under the age of 5 that die each year. Never look ahead, just use up all the resources and see what happens. The world will come to our aid –they always have!! And we have the "Ark of the Covenant", produce the world's best coffee, and beat the Italians in battle in 1896 – What else could anyone aspire to? Pure paradise.

It does make us realise how important the natural forest areas are on this farm, and how good it was that Abebech and Tibebu had protected them over the years. We now have the opportunity to help enhance these protection efforts and it is a major focus for our final six months.

As we drove back into Ambo, we had more excitement for the day. Coming up the road was a dancing, throbbing, yelling mob of about 200 young men and boys. We stopped and watched them pass. In the middle were two men, looking very miserable, carrying a large portion of a beef carcass on a stick between them and flanked by a couple of policemen. Moti told us that it was a "name and shame" activity which would have been ordered by the court. The men likely had stolen the oxen and, as they were cutting it up, they were sprung by the police. They are then made to carry the carcass around the town while the locals prance and dance about yelling "thief, thief, liar, liar". To be honest, we thought that sort of punishment finished in the 1700's when people were put in the stocks in the town square and others threw tomatoes – but there you are, still all the go in this country!!!

Home for a late lunch and in the afternoon, I helped prepare the maize that we are to take to Ambo early tomorrow. We had left Moti in Ambo, and he was to get the bus to Holeta to visit his family for the weekend. He certainly needed the break but would only leave the farm if he knew that Pip and I would be here. He had left instructions for Getun and the foreman Jagoma as to what the labourers should be assigned to for today and tomorrow. Another good sign for us. Previously Tibebu would disappear for days and leave no direction!!

Chapter 15

July 2015 - Part of the Woodwork

EE.59: At the market – unrest

One slightly intimidating feature of Ethiopia is that public toilets simply do not exist and there is certainly no such thing as a "Maccas" just around the corner where you can duck in and use the facilities if needed. We learnt to always have toilet paper with us. Almost always!!

Day 526

Up at 6.30am as last night Getun had said he wanted to leave for Ambo with a load of maize at 'sanyo one time". This meant tomorrow (sanyo in Afar Oromo language) we leave at 7.00am (being 1am Ethiopian time as the clock had started at 6am) and the "time" clarifying the o'clock as distinct from the number of people, or bags of maize.

I did not help put the bags in the ute as they are about 100kg each and I am trying to save my back for a few days. We headed off with Getun driving (he is actually getting quite proficient, and an accident is no longer inevitable) and the shop keeper Aberash to help negotiations.

As we came into Ambo, even though it was before 8.00am, thousands of young people were massing outside a large Government office all wearing bright orange t-shirts with a face on the front. I suspected it is the face of the opposition candidate in this area who was expected to be elected in the recent elections but was beaten comprehensively by the Government candidate – although any number of people claim to have seen the ballot papers being destroyed on the night of the election. This all made me a bit nervous, particularly as I then realised there were police and heavily armed soldiers all the way up the main street on both sides. Although we got to the top of the town, I doubted we would get back. We pulled off the main road onto a side street where we expected to deliver some of the maize, but the stall was unoccupied and there was a man lying unconscious or dead on the ground right beside the stall. We parked about 10 meters further up the street and Getun tried to ring the lady stallholder about selling the maize. People began to mass around the body, including a couple of soldiers and police from the main road. After about 20 minutes a police 4wd vehicle arrived and they put the man in the back and headed off, and the onlooking crowd dispersed.

I am not sure if it was the tension, or the early morning start, or the beginnings of a bout of gastric, but by now I was feeling as though there was no way I was going to get home without requiring a toilet or hole to squat over. I was eyeing off the low bush on the side of the street but for one of the first times since we have been in Ethiopia I had not a scrap of paper with me or in the ute. Things were not looking pretty but I could hang on a little longer – or could I??

Eventually the merchant lady appeared, and we sold all the six bags of maize to her at a good price. I had lost interest in the price with just one thing now on my mind. As we manoeuvred the ute to go back out onto the main road, the police and army closed the road. The significant pain in my lower abdomen grew a little worse!!! Here there was no little bush on the edge of the street, just swarms of people. A minute later a police vehicle went past with sirens blaring and following behind were marathon runners. I realised that this could take a while, so I told Getun I was going to walk down to the main hotel, and he should meet me there when the road reopened. Hopefully, I would make it.

I considered heading out onto the road and running with the marathon men and women to hasten my trip to the hotel but thought a large, old, white man running down the main street hand on backside, accompanying small, sleek Ethiopian marathon runners, may attract some unwanted attention!! The movie Chariots of Fire was foremost in my mind. Could I beat them all down the road? The music from the movie drummed in my ears.

The hotel toilets are easily accessible but the first three had no paper and I was wondering if it would be possible to dissect the cardboard of an empty toilet roll for wiping purposes??? Fortunately, I did not have the put the experiment to the test as the fourth had a few sheets left on a roll.

Greatly relieved, I walked back up to Getun in the ute and soon after the road was reopened, and we headed off. There were more and more orange t-shirts and security in the main street, but we got past the Government building which by now had many thousands hanging about, and we headed for Guder. Hopefully I will eventually find out what it was about, but I suspect the students decided to have a rally because they knew the marathon was coming through town and would therefore attract attention.

EE.60: Life in the hut

When we had first arrived at the farm, the cashier Kidist spent most of her time carting a little boy around strapped to her back. As the months progressed the little boy started walking about and more and more often ventured towards our hut. It was not long before we were acting as doting grandparents to the dear little Lami. He was spending many hours a day with either Pip or me and he virtually never let us out of his sight.

Most mornings as we opened our bedroom door, he would be waiting for us - sitting on the swing we had setup in our courtyard. Then it was time for him to count the spoons full of porridge oats into the saucepan – always with an extra spoonful for Lami. Pip spent hours and hours with him, treating him like a grownup and speaking English which he picked up in no time. Often at the end of the day, he would drop off to sleep on my knee at the computer, and I would carry him back to his hut and hand him over to his mother Kidist.

Day 538

As always Pip had her little shadow Lami all day and they did the washing and lots of baking. I love listening to the exchanges – “Lets read a story about a little sheep, Lami” “No Peepee, lets read a story about a big sheep” “why not a little sheep?” “Why not a big sheep?” “OK, we will read a story about a sheep”, “OK, about a sheep” – and so it goes on hour after hour, but there is nothing about him that annoys us. How lucky we are.

After lunch we had the help of guard Abara to cut more fence posts for the fence we are planning. We have raised funds from family and friends in Australia and the fence is designed to keep the neighbours animals from grazing the precious strip of native vegetation which has been preserved through the farm. I am still undecided what to do about strainer posts. It is annoying me. If only we could go and buy treated posts or decent concrete posts. I have seen too many home-made concrete posts fail over the years, but of course the termites will eventually destroy them if we use eucalypt.

I made sheep kidneys in gravy on rice for dinner. It would have been really nice if the rice in this country was not so horrid. How can rice not be rice that is good to eat, you ask? Answer - when it tastes more like sawdust than rice. Decent rice is getting up there with decent cheese, swimming in the sea, KFC, fresh crisp apples, sausages and Weetbix that I am looking forward to most in six months’ time - not to mention family and friends of course!! Ah, and my sheepdog Yambo, and the smell of a shearing shed would be good too.

EE.61: Merchants

At the heart of much of the hardship and corruption that occurs in the rural countryside is the merchants. There are many of them in each village. Always large, fat greedy men who live not in huts, but in houses behind large steel gates and even with glass in their windows. With millions of small farmers all growing a tiny amount of produce, it does need to be aggregated together so that it can be taken and sold properly in the larger towns or on Addis. Eventually I hope that cooperatives are formed where the small farmers can help themselves and not rely on these fat cats.

Day 564

We were out of bed in good time as I wanted to be ready with the tractor and some tools as I was expecting the hole digging men for the new fence to come and cart the concrete strainer posts to the new fence line. As we decided to have the concrete posts made at the farm, I have engraved into each post the names of the friends and family in Australia who had donated money. Their generosity will be forever imbedded in a fence in a remote area of Ethiopia. Truly grass roots aid to the developing world.

Working on the new boundary fence with the hand poured and individually identified posts noting the family and friends who donated from Australia to the project. Note the inscription for my brother ROB on this particular post.

The workers did not appear until about 10.00am and by then Pip and I had finished our morning coffee. I had spent a fair bit of time with Moti particularly explaining to him why I am unhappy with a deal he discussed with a merchant to buy some large eucalyptus trees.

My opinion of merchants around here is very low. The guy that has bought the cypress trees for 46,000 birr (A$2300) with a signed contract, came to tell Moti that the price of timber in Addis has gone down and now he can only pay 30,000 birr (A$1500), after he has already taken nearly all the good trees. Moti and his other managers thought that they needed to negotiate a settlement and offered him some more trees (Casuarinas and Grevillia's) if he paid the 46,000. I am convinced it is a total sham and when I put all the arguments to him, Moti decided that he was being conned and called off the deal. I could rave on for hours about the merchants. They are just total, corrupt dregs of society destroying the country in my opinion, and now it looks like this issue will end up in the local court because I am insisting that someone stands up to them. Fortunately, Moti stopped them taking trees as they had not made their final payment by the agreed date, so we have some good trees left plus the smaller ones that the merchant had not got around to cutting. This merchant initially tendered a price well above anyone else and I am sure that he always intended to walk away once he had taken the best trees; a sleaze bag like all his fat merchant mates.

Moti is doubtful how we will go in court (which is free to appear, a bit like VCAT) as he says the merchant will probably just pay off the magistrate. So, it is all very sad. Tibebu is matey with the main merchants in Ambo and Guder, so I have advised Moti to listen to his advice but to not necessarily take the advice and scrutinize it very closely!!

It is a terrible situation - if for example the merchant gets a lady who has carried a bag of teff thirty kms to town and it should be worth six hundred birr they will not be content to pay her four hundred for it. No, these buggers know that she cannot carry it back home so will offer one hundred birr. Absolutely no social conscience knowing that she will go home and not be able to feed her family for the next month and some of her children will probably die as a result.

I took the tractor and trailer down to the new cow shed where the concrete posts have all been poured and organized more fencing.

Meanwhile Pip had done our weekly wash of sheets and towels as the power was on most the morning so she could heat up water. Later in the day, she made biscuits and cakes with the help of Lami. He knows where everything is on our shelves and runs back and forth suggesting different things that Pip might like to put in the biscuits Coconut? Sultanas? peanuts and so on.

Another cow calved today, and I suspect another in a couple of days, so the calving flush is on us as expected. We sure need to get the new cow shed finished as the dairy herd has built up beautifully and we are running out of space for the cows with baby calves.

EE.62: Per–diem, bribe or coercion:

We have discovered so many Ethiopian cultural traditions in our farming activities and in the local community so I suppose we should expect similar traditions to exist in the Ethiopian business world.

Day 565

In the morning, we went over and put the stay rails into the strainer assemblies of our new fence. We then met with Moti and Getun to look over the cow production records. Not much to see as very few cows have been milked recently but with all the new calving in the last few days, we will have much more to review next month. We have a cow with a very sore, infected foot that we were not told about for the first couple of weeks and by then she was not eating much and was not able to walk around to graze. We eventually got her onto the correct antibiotic and now she is starting to recover.

We have introduced a chart on which milk production for each cow is plotted every Thursday. That way we can look at a simple chart and see how each cow is tracking over time. This particular cow in the last two weeks has dropped from eight litres a day to four litres and, as we pointed out to Moti and Getun, her production will not go up again even when the foot fully recovers.

We then multiplied the lost four litres a day by twenty weeks left in her lactation by thirteen birr (65 cents) per litre and showed them how the loss from the sore foot was more than seven thousand birr (A$350) – enough to pay a manager for two months, or a day labourer for almost a year. To say that their jaws hit the floor would be an understatement.

Should we have reacted quicker we asked Getun and Moti? Should we have spent two hundred birr (A$10) getting the vet if we were not sure if we could treat the foot properly? The vet (paid by the Government) is meant to be free, but he will only really come if he is sure of a per-diem as he calls it. Not really a per-diem at all, as he is already paid. It is wholly and solely a bribe!!!

We hear of this so often in Addis. If you want to organize a conference (say on maternal health or what-ever) you need to pay all the attendees a per-diem, or no one will come. Again, of course it is not a per-diem as they are already being paid by their organizations (often NGO's), but they need to be bribed to come. This is a supremely difficult country in which to do business. Where the future lies for it in a business sense, I have no idea. I guess it is typical of Africa as a whole, but maybe there are extremes of difficulty, and I am guessing Ethiopia might push those extremes.

In the afternoon we were delighted that Nikhil, a young American Peace Corps volunteer based in the local village of Guder, brought his parents up to visit us. They are visiting from America and have stayed a few days with Nikhil in Guder, after a couple of days in Addis at the Sheraton. His Dad is a doctor in the US. We have seen photos of their home in the US, and it is fair to say that Nikhil's abode in Guder is close to the opposite end of the spectrum to the lovely house where they live in Virginia.

They are of African/Indian descent and more than twenty years ago they spent a year or two in Zambia establishing a children's hospital. They ended up getting thrown out of Zambia, as despite raising 60 million US dollars and getting the children's hospital built, they had refused to pay kickbacks to the President's wife. Another example of African business etiquette!! They are really nice people, and we thoroughly enjoyed the afternoon chatting with them over cups of tea and Pip's home-made chocolate cake and biscuits before they headed back to Guder.

The electricity was on tonight so for dinner I roasted the chicken that we had bought live from a guy out on the road on Monday.

EE.63: New Year:

By now we were getting towards the end of our assignment, and we were happy to head into Addis every couple of weeks instead of every two months which had been the case earlier in our assignment. Also, with production now really cranking up at the farm there was always extra produce to be taken to the head office to be used in the orphanage, or to be sold on behalf of the farm. This is the ideal time to be delivering produce as in the Ethiopian calendar, the New Year begins on September 11th, so this weekend is a time of huge celebration.

Day 572

There was lots of packing of the farm ute and our Landcruiser as we were to take both vehicles into Addis. The Hilux has still not been starting properly and requires new timing belts and glow plugs. Lots of butter, cheese and oil was loaded into the ute and forty dozen eggs into our vehicle - along with all our gear including a basket of washing to put in the washing machine at the home of Consul General Stuart and wife Linda where we are staying. We are so lucky to have Stuart and Linda as friends and are ever gratefully for their hospitality. It must also be saving AVI a significant amount by not having to put us up in a hotel.

We are constantly trying to come to grips with how many people we were taking to Guder, Ambo, and Addis. I understood that I had three to take to Addis, so I packed with three spare seats. By the time I left I had 7 in with me, and I think Pip had even more, as suddenly bags of teff and corn were loaded onto her ute and lots of "presents" from the country to take to the city relatives who are staff at the head office. Bags of charcoal in the back, small bags of butter and cheese and even a live chook which Pip did not know she had inside the ute until she got to Addis three hours later.

It is New Year this weekend and this is the equal biggest holiday, family, and feast time of the year - along with Easter. By the time she got to Guder, Pip had unloaded most of her passengers and the local produce and the rest she deposited in Ambo. I dropped a few in Guder but gained some others, so still a squeeze to Ambo when everyone got out except one of farm guard Angussa's daughters, who I took through to Addis. We still rarely know what is going on and just cope as best we can.

The trip to Addis was amazing with every town and village having thousands of people selling chooks and sheep ready to slaughter in the next couple of days. Also, there were people selling the brush "candles" about 3 meters long ready to light up on New Year's Eve. Maybe some of the truck drivers had started to celebrate a little too early as there were three trucks tipped on their sides beside the road with their loads spilled out at different spots along the way. At one stage I had to drive around a policeman who stepped out onto the road to try and wave me down for some alleged offence! My passenger was pretty amazed that I did not stop, but I am not really interested in paying them a bribe so they can buy an extra sheep for New Year for their family. They do not have vehicles so there is no risk of being chased!!

Sheep are available for purchase from the side of the road all over Addis Ababa and the rest of Ethiopia at festival time.

We spent a while at the Head Office delivering all the goods and meeting with managers and then headed to Stuart and Linda's house where we will spend the next couple of nights. Linda has taken a three-month posting to Kenya processing refugees, so she will not be there this visit. Stuart arrived home from work at the Embassy and then he took us to the Ambassador's Residence for a reception. There were about 30 Aussies present and we all had a good chat with the DFAT Deputy Secretary for Africa who is visiting from Canberra. Also, we all get a chance to catch up with each other's activities. It is always good to see Ambassador Mark and his wife Anaise, as we get on very well.

The traffic in and around Addis is terrible with everyone on the move for holidays, rain every afternoon, many spots where the roads have completely fallen apart and people and animals everywhere. There are literally thousands of sheep in little mobs all over the city, waiting to be purchased and slaughtered for the New Year, and frequently they are on the road or blocking a lane or two.

EE.64: Services of the "night":

Prostitution is obviously a major business in Ethiopia, but it is not too prominent in the areas where we spent most of our time. There is one area of Addis where it seems every door in the street leads into a brothel and this area has somehow acquired a search light which at night is often beaming rays of light into the night. Maybe men are attracted to the light like insects!! Also, late in the evening women in pairs often come and sit in the lobby of the hotel – presumable waiting for business.

Apart from the potential of local clients, Addis has thousands of expats in residence, with over one hundred embassies (many servicing all of east Africa), the head office of the African Union, and literally thousands of Non-Government aid organizations from tiny groups assisting marginalized communities to large groups like Save the Children and World Vision who each have thousands of employees in the country.

Day 573

Addis is really busy again as everyone is preparing the holiday weekend. At this time of year there are many road accidents as people from the country sometimes drive into Addis and have no idea how to cope. Demiss warned us to be very careful. Pip and I walked to the nearest supermarket in the morning and got a few things for lunch and dinner. We also stopped at a drug store and got some eyedrops for Pip as she has a swelling under one eye.

We stopped for coffee in a place that looked a bit seedy and instead of menus on the table they had laminated sheets on which there were photos of about twenty women. No doubt the women were "available" upstairs and the "menu" with photos also had the prices they charged for different services. Unfortunately, I could not read the Amharic to tell me what I could get but there was no doubt "full service" was available for 40 birr ($2.50) and I suspect thrown in you could get a dose of HIV or an STD. There was something you could get for 7 birr from one of the ladies, but sadly that remains a mystery!

Despite the traffic we found our way to the Hilton for Pip to get a haircut. Again, I was so glad that in recent years we have never, ever relied on a Navaid to get us around anywhere in the world, and so we have been able to maintain a common-sense approach to navigation. It would be impossible to drive in Addis if you had lost that skill. (Take note next generation!)

EE.65: Biribisa – a simple but good life

> *Our day-to-day life was simple and uncomplicated, and I think we were starting to understand the benefit of just living for today and never thinking for tomorrow. That is especially true as we begin to think of the future - having to leave this place and people with whom we had so fallen in love.*

Day 589

Pip and I think it would be better to sell the male calves as soon as possible after weaning at about eight weeks, as looking ahead a year or two and calculating if money can be made by keeping them is way beyond Moti, let alone Getun. It will not be a matter of lack of academic ability, just an ingrained inability to look into the future. Now, with almost thirty female cattle in the herd, plus eight working oxen to be fed, there will not be any spare feed about unless the cropping program is reduced. It is time for us to plan a course to convince the management of selling the male calves early.

I spent ages sitting down the paddock hooked onto the internet trying to research the options for flights back to Australia in February as we have to submit three quotes to AVI. They then pay us the amount for the best of those options, and we can then spend the money however we want getting back to Australia. I thought I had all the information saved after a very laborious effort with constant internet drop outs, but when I got back to the hut, none of it was saved. Damn.

Late in the afternoon we went for a lovely long walk down through the Birbirsa "village" area and through the cropping areas to the river well downstream from our farm. We passed through an area of eucalypts which is owned on shares by Tibebu and were amazed to see many of the trees already three meters high, even though it was harvested only nine months ago. Tibebu also has an acre or so on his farm behind the plantation which now has many good-sized pumpkins. We followed the river a bit further down and then returned to the bottom end of the Birbirsa village area, about one and a half kilometres down the road from the farm. It was a lovely warm afternoon with the setting sun and the usual swarm of children that join us down in that area and back onto the road. Everyone is in great spirits for the Meskel long weekend and walking back up the road many locals were out and about, and we shook hundreds of hands by the time we got home.

What is called the village area is an area of lovely grass tracks dividing up scores of half hectare lots which are owned by the local people. They mostly have one or two tukuls in which huge families are raised along with goats, sheep, and a few cattle. Each block is surrounded by shrubs and eucalypts along the dividing lines which tend to make a live fence and gives the whole area a beautiful, serene look. Many of the people who work at the farm live in this area, so we are well recognized. As we walk past, the women often come to their gate surrounded by their two or three youngest children, and any of the children more than three years of age have the confidence to run out and surround us and add to the crowd. They now all constantly call "Sime, Pipee" as everyone knows our names. At different intersections on the grassy tracks, the crowd of children tends to change, as some drop off at the end of their range of territory and others join in as we get onto "their" terrain.

A beautiful evening with an almost full moon and no rain today as the "rainy season" has finished, so we lit the campfire for the first time in ages. Cooked popcorn on the fire which we shared with the duty guards and Pip also roasted our next lot of coffee beans we had recently purchased. Some mutton chops on the fire and I cooked chips to go with the mixed salad of lettuce from Pip's garden and purchased tomatoes and avocados. The avos have become delicious again. We always have them on hand, but they have not been particularly nice for the last few months.

EE.66: An assignment to match our skill set:

In Australia, Pip and I had spent our lifetimes learning and picking up more and more skills in the agricultural areas. Our farms have run various enterprises, and being quite isolated, we've become quite resourceful. We were continually amazed how much of our knowledge base we had the opportunity to use during our assignment.

Day 596

The weather has dried up really quickly at the end of the wet season and the days are all pretty warm (27 degrees), with a breeze most days which seems really drying. We washed our hair in some bowls of warm water to save the extra time washing hair in the cold shower. I have a quick shave in the cold shower every night - which is not pleasant but keeps the whiskers at bay.

The foreman Jagoma came to work today to organize the day labourers as usual. Moti called him into his office and when we later caught up with Moti, he told us that he had told Jagoma that he must do some work at the same time as he supervises the labourers. All of the time that Lechisa was here last year we argued with him that Jagoma was useless just sitting under a tree all day watching the workers. To my surprise Lechisa used to defend him and say that he was important, keeping an eye on things and reporting issues to him from around the cropping program. I did not accept it.

Then when Lechisa left and Tibebu was hardly ever here, there was a management vacuum for the cropping, and I was most surprised and pleased to see Jagoma lift his game and show quite a lot of initiative and start taking some management responsibilities. When Moti started as the manager, it was clear that Jagoma just started sitting again and I tried to get Moti to give him more responsibility. Pip and I even suggested at the management review a few months ago that he should be asked to sit in on management meetings and he should have a major role helping Moti (who has no agricultural experience) with the cropping program. Moti has not recognized any benefit from him and says that he is not even keeping the workers active, so this morning he was told to work or leave, and he chose to leave!! Moti says he will now meet the day labourers first thing at 8.00am each morning and allocate who goes where.

Time will tell if Jagoma is back on the doorstep on Monday morning or if Moti can handle the day-to-day organization of labour. Pip and I are more convinced than ever of the need to simplify the cropping program as much as possible and focus on the dairy, sugar cane and teff.

There have been two old Yamaha motorbikes in the store covered in dust which I have been keen to have a go at. Apparently, years ago they were used to travel around the different health clinics in this area that were established and run by our organization and then left to rot with no knowledge around the farm as to their mechanical status.

Storemen Abissa was at the farm today, so I got one of them out of the locked store shed and, using more than 40 years of experience trying to keep Yamaha motor bikes going, I had the bike running after an hour or so. The back tyre obviously had a problem but fortunately Pip and I had brought some "puncture repair Slime" back from England last Christmas especially for this occurrence. I put some of that in and pumped up the tyres and all is good. I rode around the farmyard for a start and then Pip had a ride around.

It is almost unheard of for a woman to drive a car in this country so for a woman to be able to ride a motorbike is unthinkable. Needless to say, the spectator's eyes were again out on stalks as they have been so often during the last year and a half. We often marvel at how lucky our assignment has been and how well matched our skills, experience and attitude are to the needs of the farm. We could write a list a mile long of skill set matches from Pip's animal husbandry and eucalyptus plantation skills to my bookkeeping and (admittedly limited) mechanical skills and a myriad of things in between. If the motor bike were anything but a Yamaha, I probably would have struggled to get it going, so the luck continues with us.

As every vehicle is ridiculously priced in this country, I will try to encourage the management to sell the bikes and use the money for something else – like repairs to the Hilux ute. They should be worth a heap more if I can have both running properly. Nobody at the farm has ever ridden a bike of any kind, let alone have a motorbike license, so there is no point them sitting in the shed here.

EE.67: Farm meeting in the Managers office

At this time we had certainly started to notice a definite shift in outlook across the whole farm. Almost all areas were now starting to look productive, and we felt that a new emotion of PRIDE was starting to develop amongst the staff and managers. When we had first arrived at the farm there was rubbish everywhere, including thousands of used syringes with needles intact that had been thrown out on the ground behind the medical clinic. Pip and I did not say anything about it but slowly just started cleaning it up ourselves. Eventually Tibebu and others started to notice and, when at last someone mentioned it, we took the opportunity to start introducing the concept of pride. Gradually we felt that keeping the place neat and tidy became second nature to the staff.

Day 599

Soon after breakfast we were told that our presence was required in Moti's office, so we left Lami playing in our hut as he is pretty responsible now and we can see and hear from Moti's office. Moti has assembled all the managers for the first of what we hope will be regular meetings. It turned out to be the most frank and productive meeting we have ever had in the manager's office here.

Moti had agreed that the managers needed to be included far more in the overall financial management of the farm and so I printed out a copy for each of them with the projected yearly profit and loss statement for each enterprise and handed them around. Currently it shows a slight profit for the year overall, but also it gave them a good chance to see how the managers' salaries are a substantial part of the expenditure in each enterprise. Discussion went around the different enterprises and who could make what improvements.

To our surprise, Getun then asked what else Pip and I thought could be done. I suspect he was waiting for the usual "work harder, too much staff salaries etc" and I could see everyone's mood lightening up as I talked enthusiastically about all the things that are in place that will increase profitability if they are well managed by the people in this room. Far more cows being milked, more milk per cow, no cattle deaths so more to sell, far more grass due to the fence, better grazing management, concentrating on the profitable crops, increasing the sales from the shop, getting organized to buy far more Niger Seed for processing this harvest time (because we now know our cash availability situation), beginning sugar sales early next year and so on.

Most of the managers joined the discussion enthusiastically except purchasing officer Abara who suddenly made a speech about how he already has too much to do! Pip and I started to laugh, and we rather suspect others joined in. He says he is now the purchasing officer (no more than twenty purchases are made by the farm each month), accountant (no accounting done at all) legal officer (no legal matters ever happen), gatherer of records (head office asked him how many cows the farm had a couple of weeks ago and he had no idea!!) and now we expect him to be in charge of maintenance as well. We think he got a good dressing down by Moti and some others, but this was not interpreted to us. Most of the discussions are in Amharic. I understand some of it, and Moti translates when important.

Finally, we had a good discussion about increasing sales from the shop. They even agreed that a blackboard with the products available each day out on the road would be a good idea. We discovered that the shop used to sell far more products and all the milk, cheese and butter was sold directly from the farm as well as some grains and even clothes at one stage. This was complete news to us and apparently had only stopped when the management just got too lazy to make sure the shop was stocked with items to sell. Who knows, maybe if we keep pushing, another of our dreams will be realised.

The meeting ended with Getun leading the others in some very heartfelt thanks to Pip and me for our continued efforts - and how they could not believe how we just got on and built the fence ourselves and they knew they should have helped but never quite got around to it. They have promised to come and help (and learn) as we do the last little extra bit that we are still planning. Moti had done an excellent job of running the meeting, letting the others have their say and also translating the important bits for us.

Chapter 16

Oct 2015 - Wheels of Change

EE.68: Medical Emergencies

In Australia, we have always lived quite a distance from a hospital and so have been pretty much self-sufficient when it comes to illness and injury. Over the years, we have maintained a reasonable level of first aid training and refreshed in that regard before we commenced our assignment. Also, dealing with animals almost daily, we frequently must perform minor operations on them, so we felt pretty well equipped to look after ourselves in Ethiopia. We had not realised how often our first aid training would be required in the Birbirsa community.

Day 612

Lots of labourers turned up today for potato digging, so plenty of action. Up and down the rows the oxen went, turning up the soil full of potatoes and lots of ladies behind with bags for the potatoes and carrying them to the tractor and trailer. Up at the shed, we unloaded a big heap from the trailer, ready for the merchant to come in a day or two and bag them up, weigh each bag and take them to Addis.

Bagging up potatoes

I just got to lay down for my afternoon zizz and there was a racket outside the door. A crowd of workers and others had appeared with one of the cowboys, Gardessa, who had somehow opened up a great long vertical gash between his ankle and Achilles tendon. It definitely needed stitching, and Pip and I figured he would also need a tetanus shot. We probably should have set to and started to wash it properly as Pip suggested but, as it had stopped bleeding and I did not want to get it started again, I thought I could whip him into Landcruiser and get him down to the new Guder hospital where he would be getting proper medical attention within twenty minutes. We loaded him into the cruiser with newspaper on the floor to catch the blood, and I also took our purchasing officer, Abara, to do the negotiating, as he speaks a tiny bit of English so could keep me in the loop.

It was an interesting experience at the hospital where most of the worst of Ethiopian culture was on display, but I had a most fascinating time watching it all. First to pay ten birr (50 cents) for an appointment. Then to the door of what must be the emergency treatment doctor and room. It was in a courtyard with about eight doctors' rooms around, but in the two hours that we waited most of the other doctors did nothing at all, most of them were arrogantly playing on their mobile phones whilst patients waited outside their open doors. Others strutted about with stethoscopes around their necks thinking they were Oh so important.

I thought we would be here for at least two days as there were about twenty people on the seats outside the door of the emergency room. Eventually a sick person came out with about five family members from inside the emergency room and another fifteen family members then got up from the "waiting room" seats and headed off. That sick person must have brought at least twenty with them as their support act. Now we were down to about half a dozen waiting, including our boy with the gash. As the last group came out, all the half dozen next patients pushed into the room at once trying to make sure they were the next one treated. I stuck my head in to see the chaos and fortunately Gardessa and Abara were pushed into a second treatment room. Even then an old guy followed them in and thought he could get up on the table first, but he was hunted out to the waiting area. The doctor gave Abara a list of items he would need to patch up Gardessa, and we went around to another window in the building and purchased a scalpel, needle and thread, antiseptic, bandages, and antibiotic capsules (sixty of them!), I paid the one hundred birr (A$5), and we went back to the doctor. I was tempted to stay and see the action but maybe I am not Ethiopian enough yet, and I really feel it is not a spectator sport. Abara did stay to watch.

All the time more people coming for emergency treatment keep pushing the door open into the doctors consulting room and even the treatment room trying to be next cab off the rank. I went for a walk around the hospital, which has only been open about six months so is very modern by Ethiopian standards. Looking at the state of repair, you could easily think it was about twenty years old but there were rooms that had signs for prenatal, delivery, ward, x-ray, and more. They could have just been signs or maybe there actually was something in each room to back up the sign, it was hard to tell!!

By the time I got back, Gardessa was patched up, so we took him back to the Landcruiser. He still had bare feet covered with dirt and blood but presumably the wound area had been cleaned up. The hospital drug store had no tetanus vaccine, but the doctor had written down what we needed to purchase in Guder, so we drove back the couple of kilometres into town centre and went to two different clinics but neither had any tetanus. I was adamant that he needed tetanus today and so we drove to the larger town of Ambo and a drug store that I know near a big clinic. Sure enough, they had some tetanus vaccine for which I paid eighty birr (A$4). Then Abara took Gardessa into the clinic next door to get the vaccine administered. I had been wondering if I was confident enough if required to do that job!!!

I left them there as it looked like a long wait, and I was getting worried about driving home before dark. I knew there were some American doctors coming to Ambo today and I thought they may be at their hotel, so I drove to the hotel. They had "checked in" but had gone to Guder hospital to check it out ready to start tomorrow doing a weeklong clinic.

I drove back to pick up Gardessa and Abara, and at one stage had to squeal to a halt when a six- or eight-year-old boy inexplicably ran straight across the road in front of me. The thousands of people in the main street of Ambo were silent for a moment, I am sure. I missed him by a good margin but left plenty of rubber (or was it canvas) on the roadway in doing so.

With the injection having been administered, we headed for the farm, dropping Gardessa as near as possible to his house and arriving home just before dark.

Back from my wash under the irrigation bore and I was just getting dressed and another rap tap tap on the door and another crowd outside. Our mate, young fourteen-year-old Hordofa had been bashed with a rock by some boys and had a huge hole above his eyebrow. It must have been a fearful blow with a sharp rock as you could have put your finger in the hole and with the swelling it looked a good centimetre deep. Out on the veranda we washed it as well as possible with our filtered water, and, as it was dark, we then needed to take him into the light of our hut for a decent examination.

As Pip and I came and went into our other room to get bandages etc we had to push our way through a dozen or so guards and others from around the farm and community who had come in "to help." He badly needs to be stitched but there is no way to get him to a clinic tonight so we patched him up as best we could and sent him home with his Mum and some pain killers. Hopefully she will take him to a clinic in Guder tomorrow. Pip said he also had a good lump on the side of his head where he had also been whacked but I did not see that. Hopefully, he does not have concussion, but he seemed to be focussing okay.

I can't believe that someone would not talk to the police about it, but I would be pretty sure that will not happen in this society. I imagine he is quite often picked on as he is quite artistic and much better educated than most of the boys around here. He will surely carry a scar for the rest of his life.

Eventually, dinner of lamb steaks after an eventful day. Just yesterday I was thinking how quiet it had been in recent weeks for serious injuries, deaths, attempted suicides, etc.

EE.69: Mechanisation – should we or should we not?

Before arriving in Ethiopia, there is no way we expected to become completely conversant with how to plough a field with oxen. After all, our assignment was to introduce modern technology and innovation. Actually, it took us no time at all to realize that attempting to mechanize the activities of the farm would do way more harm than good.

Day 613

A busy day for the farm labourers again as it was another potato digging day and the merchant came in his truck at about 9.00am with a team of about ten young men. The farm workers had started at 7.00am for more digging of potatoes. The merchants' workers then picked up the potatoes out of a big heap in the shed and put them into bags which were then sewn, weighed, and loaded onto his truck. He headed off to Addis Ababa mid-afternoon with about five ton of potatoes on his truck. There are still some potatoes to harvest as the oxen will next plough across the paddock in another direction and that will turn up quite a few more.

We had an interesting discussion with Tibebu the other day about farm mechanization in Ethiopia and were left to wonder how it will ever become widespread. The farm was given a tractor and some basic farm machinery (a disc plough, a trailer, and a set of offset discs) about fifteen years ago. None of it has ever had any maintenance and virtually any breakdown will mean that the machinery will be left forever to rot. We rarely see a tractor anywhere in the country and the few that we do see are often parked in a state of disrepair. Fortunately, this farm was given a Massey Ferguson and not an unreliable Russian or Chinese brand which we have seen occasionally. It has done a remarkable job continuing to operate.

The set of offset discs (also Massey Ferguson) have always lived outside and so now the bearings are beginning to fail as the rain works its way in year after year. The consequence is that we need to replace just 4 bearings, and the cost of the bearings alone is more than the value of the whole potato crop. Here lies the big dilemma with mechanization. The income from crops is traditional, but the cost of spare parts for machinery is definitely modern, and it is impossible for the two to match. We will get our son Richard to bring the bearings from Australia as they are only half the cost in Australia than they are in Ethiopia, and we will pay for them. Even at the Australian rate, there is no way the farm could pay, so the discs would be just left to rot, and the oxen would have more work to do in the future. It is hard to see how that will ever change. To return to the biblical reference, is the giving of the bearings "giving the fish" or teaching "how to fish"?

On the face of it, certainly it looks like giving of the fish, but if that does not occur, how do they ever learn the ways of modern agriculture? Increased agricultural output and more efficient production methods mean a better level of nutrition, and better farm income which ultimately leads to better education, and only through education will the population explosion be controlled and give the country hope of long-term survival. More food for thought!!

Eventually the tractor is sure to have a breakdown and repairs even for a minor issue will far exceed the total output from the farm, so I guess it too will get parked eventually.

No vehicle or equipment is ever sold as far as I can tell, as almost everything of any value has been donated to the different charity organizations in the country. The paperwork (and maybe legalities) seems to exclude sales of vehicles and machinery. There are at least six broken down vehicles (including an ambulance and Toyota four-wheel drives) in the sheds at our head office in Addis. I can understand they may be too expensive to fix, but I argue that surely then you would sell them for spare parts and use the money for other repairs. But that seems impossible, so they continue to block up the shed space, and the working vehicles live outside. I doubt the organization has paid for any of them and that creates this odd "give me" mentality so rife in Ethiopia – and to which we may be adding!

Whilst generally the people are very careful about looking after themselves, their axes, and their hoes, exactly the opposite applies to looking after anything more sophisticated. They just have no affinity for machinery so anything mechanical receives no care or maintenance and is treated appallingly. If there is a single large rock in a paddock, the tractor and plough just have to be driven right over the top. If there is a deep trench across a track, the tractor and trailer just have to be driven across it in road gear, regardless of the impact on the vehicle.

In the interests of mechanization I designed and built this "implement" to improve the accuracy of seed planting. Made from water bottles and garden hose.

We have been very careful in formulating our recommendations and experiments not to base them at all on mechanization. It has certainly added to the interest of the job for us. I still believe a pair of well-trained oxen would be a great addition to a vegetable farm at Werribee. However, I have no intention of leading that foray!! Anyway, where would I get a man to plod up and down behind a team of oxen in the howling wind and sleet of a southern Victoria winter? I suspect the farmers in Australia may prefer their air-conditioned tractor cabins. Actually, you probably could find the operator in Footscray, as I will bet that virtually every male Ethiopian that has immigrated to Australia knows how to drive a team of ox. It is something they learn at a very early age

.

EE.70: Patience and love

During our pre-departure briefings in Australia, we had been told that almost anywhere in Africa if you want any sort of peace and quiet, you should establish early on that the local people are not welcome in your personal house or hut – otherwise they will be there day and night – as is the tradition in their cultures. Throughout our stay we maintained our bedroom hut as our private area but it did not take long before our other room was pretty much a free for all. When we were not about, we would leave the door closed, and if away for a day or more we did lock the door.

Otherwise, managers would come and go borrowing things and children appeared constantly – and of course as time went on, Lami was there pretty much all the time as he lived with his mum Kidist on the farm about two hundred meters from us. Periodically, the farm guards would announce that too many local children were coming in off the road into the farm to visit our hut and all the children would be banished, and our hut would be empty for a day or two except for Lami. Then some of the children of the guards would start to reappear and then they would start bringing their friends and so our hut would be back to the noisy, raucous environment we knew and loved from daylight to dark.

Day 622

We had a couple of ladies picking the last of the climbing beans that we had planted on the maize stalks, and the experiment seems to be very productive, supplying excellent quality nutritious beans and simultaneously putting nitrogen into the soil for next year's crop.

I was heading down to help with the picking, but somehow I had five children trailing along with me from two to six years old. The stubborn two-year-old decided it was too far to walk so he headed off alone back up the farm with a determined march, and I was sure he would head straight out onto the road and head towards his family hut. All I could do was to trail along behind as I would not risk leaving him alone. When he got to the farm gate which leads out onto the road, fortunately one of his older brothers came in the gate and convinced him to follow to our hut – and so I spent the next hour with a gaggle of children playing with cars and blocks on the floor. All part of a day's work!!

Pip had gone down to help with the beans and reappeared at morning coffee time as they had finished harvesting the beans. She had half a dozen children for the rest of the day with different ones coming and going. I love it, but an hour is a fair bit. Pip has ten times my patience and rarely tires of the gorgeous black mites – testing their artistic talents, teaching writing, listening to their reading, teaching computer and so much more, and late today an hour of gymnastics as the kids rolled and tumbled on the bags of beans and bean-pods which are waiting to get properly dry and sorted.

Long after dark tonight, before she had even headed across the paddock to the shower room, she still had Lami on her knee reading Dr Seuss. Now three-and-a-half-year-old he is completely fluent in English. Also fluent in his native language of Afar Oromo and probably has some Amharic as well.

EE.71: School Excursion

Any chance that we had to share our knowledge with the next generation we welcomed with open arms. In a rural, farming community most of the knowledge is gained from the older generation so unless there is intervention somewhere and new ideas are considered, absolutely nothing ever changes.

Day 626

From two in the afternoon, we were looking out for a group of year ten students from the Guder high school. The school has suggested that all the Biotechnology course students should come for a visit but fortunately we convinced them that we really could not handle six hundred students! The compromise was that they should send the brightest ten percent.

They arrived at about 3.00pm having walked the seven kilometres up the road from the school with three teachers. They were here as they particularly wanted to see the biogas and the milk processing with cheese and butter making. We ended up with about sixty students, and they were truly delightful young people. Pip had made up some butcher's paper diagrams of the scientific processes, and we got Moti to talk them through the various aspects.

I had a great time with them, particularly showing them the electric fence and getting a few of the bolder boys to touch it and get a decent shock, much to everyone's delight!! One boy was obviously really bright, a real card and had good English. I complimented him on his English, and he said that he thought he was pretty good at English, until he heard me speak! Having only heard me open my mouth half a dozen times, he then proceeded to mimic my Australia accent superbly. He also wanted to know if we had a daughter as he was ready to marry her and get a ticket to Australia. If I had a daughter...hmmm ... you could do a lot worse than this boy for a son-in-law!

A group of high school students from Guder.

It was all a lot of fun, and finally they headed off at about 5.30pm to walk back to school and then presumably home. Pip and I speculated on the logistics of getting sixty Australian students out on an excursion, particularly involving a fourteen-kilometre round trip walk, no seats or chance to sit down when they got here, and not leaving school until about 2.00pm. They had no hats, sunscreen, backpacks, food, or water bottles. No puffers, first aid kits, permission notes, allergy notifications or special aids. No parents, no St Johns on standby and no toilets within a hundred kilometres. Just three enthusiastic teachers, endless smiles and politeness, a pen and paper for some - and it was as though they had just walked to the other end of the school yard.

Almost without exception (except for the proportion of girls who will get pregnant), they will end up with Ethiopian university degrees in another six or seven years and then they will begin their inevitable dream of getting to America or Australia for more study and the "rich" life.

When you think about six hundred children, just in year ten in a small town like Guder, and they are all now entering breeding age, how big does the school need to be in fifteen years? Scary stuff as these are only the ones that are going to school, and plenty of others have pulled out by year 10 to help on the farm or marry the local boy.

EE.72: Delicate issues:

With such a huge population of adolescent children we can only presume that the "facts of life" are taught openly. We were told by Tibebu that the local primary school was keen to get a toilet block with doors where the girls would have some privacy. Without doors it could be determined by onlookers if the girls were menstruating and then they were far more likely to be raped as the boys knew that pregnancy was a less likely outcome. The doors on the toilets meant that the girls could keep coming to school during their periods.

Day 627

Two of the managers are on holidays now and seem to have been away for weeks. We suspect that all the managers are catching up on the leave they are owed since store manager Abissa was refused all the weeks he thought he would take, dating back to 2011. Moti had checked the organization bylaws and found that he could only take the leave owing from the last two years.

The day gave us great hope as Moti was to be away all day, and he must have told Abissa that it was his job to manage for the day and keep the labourers productive. It is the first real sign that the managers are now taking some responsibility for the productivity of the farm, and Abissa was on the job all day and sent workers off here and there to get the jobs done.

He organized me to drive the tractor (with Getun away), and I had five girls with me carting more loads of broad beans plants back from the paddock and into the shed to dry before threshing.

You can never be sure in this country when the children are reading English, whether they actually understand the words. Late this afternoon I was bailed up by a twelve-year-old girl who was having some teff grain ground for flour by our grain grinding man. She is a great reader and insisted that I sat with her, and that she read me a new looking school book she has been given. She started with the preface - "This book has been written to give to girls as they reach the age of puberty and start menstruating so that they understand the changes to their bodies and the activities of adolescence." Then it got properly down to the serious business!! OK – not quite "John can jump, Betty can jump" as I was expecting but she did not blink an eyelid and continued reading on. After an appropriate time, I suggested that it was a lovely new book and maybe should take it to Pip and read it to her sometime. I suspect that on the whole the local people are very open about such things, and so the girl would see no reason not to read it to me. The other possibility is that even though she could read words in English, such as 'adolescence', she had no idea what it meant.

EE.73: Another day another Ethiopian Experience:

We had worked out that the farm would be much better off financially if the male calves could be sold at a young age. Like so many of the changes that we sought to implement it seemed to be that Pip and I had to implement the innovative change for the first occasion and then if (mostly) it was successful the concept may be implemented in the long term. Whether the "long term" meant just until the end of our assignment or really truly long term we could never be sure.

Day 631

Ah - what a day of trial, tribulation, and ultimate satisfaction!! Up early to get our gear packed so that we would be ready to head off to Addis as soon as the calves had been loaded into the Hilux. I had no idea how eight calves were to fit in an area that could comfortably fit four. The pen on the back of the ute was actually made for covering no more than moribund boxes of tomatoes (that could be guaranteed not to put their heads out the broken plastic windows), rather than powerful frisky calves. I feared for plenty of chaos in the loading so thought it better to take the ute down to the cow shed and then clear off for half an hour or so. When I returned, all the calves had been stuffed in (untied) and a wooden rail lashed across the back with ropes. I as started to move forward the rail snapped, and three calves tumbled out. A good start given I was in the first metre of a one-hundred-and-fifty-kilometre trip. A thicker pole was found, the calves stuffed back in, and I relaunched the escapade.

I picked up Pip and our luggage and all the rope and twine I could find, plus a Stanley knife and we set off for Addis with Pip driving and me looking endlessly in the mirrors to see what appeared over the side or out the back. Sure enough, in Guder I yelled to Pip to stop as one calf had fallen out the back. We managed to catch it with the help of the inevitable crowd that forms immediately for any drama, and lifted it back into the vehicle. I then added more ropes and ties. We could already see that the oldest but least robust calf was going down under the others but there was really nothing we could do other than press on and hope most of them got to Addis alive. We had a few more stops when heads or legs appeared out through the side of the canvas cover, and I pushed them back in and added more twine.

It was organized through Moti that when we got to Holeta we would ring Moti who would then ring the merchant and tell him an estimated time of arrival at a service station on the ring road of Addis where we would pick him up. About five kilometres from the rendezvous point, I suddenly saw a man rush out into the traffic and

I recognized him as the merchant. This is a new skill I have developed over the last two years and an essential one in Ethiopia. You must be able to recognize and notice a single familiar black face amongst a crowd of 10,000 on the footpath even though you need to concentrate to an extreme level on the road ahead. We are always picking up farm staff members in Ambo from amongst the masses of humanity on the roadside and footpath.

We picked up the merchant who was very pleased to see us and learn that we had all eight calves on board. Pip continued driving halfway across the city to the main abattoirs. I doubt that the merchant had ever seen a female driver, let alone the driver of a ute, and then with animals in the back as well. He sure had his Ethiopian episode for the day.

The abattoir has thousands of animals outside under shelters where I guess people can purchase animals and walk them over or along the road and into the abattoir to get them killed.

We cut the twines off the canopy and undid the ropes to reveal our load. The least robust calf was laying out flat, and we lifted the others out and they were covered in poo and wee but otherwise all in good condition. I had noticed the flat-out calf twitch so when the merchant pointed and said – "that one dead" I said confidently "Oh No – it alright." I dragged it out and it flopped onto the ground in a heap, but with some gently persuasion and a little bit of resting we got it to stand up – at which point Pip took a photo to prove we had delivered eight live calves (JUST). The merchant was delighted and gave us two large bottles of water (sealed), and we headed off very satisfied. We are now confident that in the future whenever there are male calves born at the farm the merchant will buy them at a young age, and the farm will be much more profitable as a result. Hopefully they will be able to deliver loads of no more than four calves in the future, whenever the ute is going to Addis.

Unloading calves at the abattoir in Addis Ababa. Note the "sturdy" stock create on the ute!!

On our way to Stuart Shaw's house, we stopped to purchase some bread. I could not help noticing a military policewoman who was stationed beside the road where we pull up at the bakery. She was a stunningly beautiful young woman with full makeup and huge gold earrings and intricately plaited hair. Dressed in her body hugging, blue camouflage military uniform with a large machine gun and massive bullet magazine draped across her knees. How's that for incongruous!!! Certainly not out of place in a fetish magazine! If you must be shot by the military, some ways could be worse than others...............................

Chapter 17

Dec 2015 - A Sea of Humanity

EE.74: Fun at the airport

Ethiopia is quite close to Europe "as the plane flies" so we took the opportunity to go to the Greek island of Corfu for the week of our Christmas period. Our return was another opportunity to share an aeroplane with hundreds of happy Ethiopians. Air travel is becoming much more common domestically and so every year the industry is becoming more sophisticated, and the local travellers are learning more of the etiquette and rules of air travel. In the years ahead, travelling with a plane load of Ethiopians will no doubt become as boring as travelling with people from the developed world.

Day 683

A flight with a plane load of Ethiopians is definitely a highlight for any traveller. You do not realise how many etiquettes we build up as seasoned travellers but most of them go out the window for Ethiopians. Apart from anything else they all look happy (is that even allowed in a plane these days?) and there is constant chatter and excitement, often with friends or relatives many seats away. Also, when you get on the plane, if a seat takes your fancy, you just sit there and settle right in. The consequence is that loading takes ages. When the folk that have been assigned those seats then arrive, a flight attendant needs be called to negotiate a settlement or to convince the incumbent that they must shift somewhere else. This all takes time but there is plenty of good cheer. The lady in the seat behind us on this flight even called out to us "where are you from" at one stage – interacting with others? Surely not allowed!!

Her companion on this flight had the loudest voice I have ever heard and the whole back half of the plane must have followed every word of her non-stop conversation with her friends and relies around her – that is if you could understand Amharic!! Most people seemed amused by the entertainment she provided. She was not deterred by it being the middle of the night and possibly a time for sleep for some passengers!!

As the plane comes in for landing, there is frequently someone who stands up and opens the overhead locker, get out their bags and start heading for the front of the plane. After all, in Addis Ababa you always want to get off the bus as quickly as possible. Needless to say, this causes some angst from the flight attendants who battle to get them seated again before the wheels touch the ground. Once the wheels do touch the ground, the action is likely to start again instantly – so much for "please stay seated until the seat belt light is switched off". Another first for me on this flight was passengers pushing past the food trolley in the aisle. Whether on foot or in a car, Ethiopians will always back themselves to squeeze through a narrow gap and I fancy the gap between a seat and a trolley was a good challenge. Plenty of determined pushers got through on this flight. There was no consideration to maybe wait until the trolleys were out of the way before heading to the toilet or going off for a chat with a friend down the aisle.

We arrived on schedule at 1.00am, into Terminal One which is the shocking old terminal used mostly for domestic flights but also by Qatar airlines. Out of the plane and into a shuttle bus and then into the Terminal. All was going well until we realised, we were jammed in the immigration room with three hundred others all pushing and shoving but making no progress towards the two open immigration windows.

Pip and I were standing about two thirds of the way back in the crowd and did not move for forty-five minutes. Not one single inch. We looked around and realised that by now we were absolutely at the back of the crowd. Only about fifty people had gone through immigration, but Ethiopians are the best and most polite pushers in the world and even though we had not moved, the one hundred or more that should have been behind us were now jammed even tighter in front of us. It was like some sort of miracle, it happened so quietly and seamlessly, so we, along with a few other foreigners, had not even noticed. We were all now right at the back.

I calculated that the two windows were taking three minutes for each person which meant that the crowd could take eight hours to clear. Fortunately, two more windows opened but by then the crowd had pushed up hard against the windows, all thrusting their passports through the window at once. Some soldiers, who had been keeping things under control, tried to get the crowd to move back but had no luck so the four immigration workers walked off the job.

We have been travelling since 8.30am yesterday and it was now 2am today and it looked like we may be there for the night sharing the floor space with a few hundred locals who incredibly were still in good spirits.!!! Our spirits and those of the few other foreigners were not exactly at the same level!! We are talking a large bare room, no toilets, food, shops, seats, water - nothing.

Somehow the crowd was convinced to back off a meter or so, but it did certainly not move at the back, so I suspect the sardines must have become stacked in even tighter up towards the front.

The immigration officers returned to the job and the processing started to get a bit quicker but there was still well over one hundred to go when we, and a few other foreigners, managed to sneak around a side queue and get processed. By now it was close to 3.00am and the luggage collection area was unbelievable. Literally hundreds of suitcases and packages had been flung off the carousel so that more could be put on. No order, just stuff everywhere.

We started to search for my backpack which was the only 'checked through' luggage we had. After a time, I noticed it on the floor in the middle of the carousel around the other side from us. As there were no officials about, I jumped up on the carousel and went around until I could jump into the middle, pick up the backpack and jump back on the carousel again and go around until I could jump off to Pip, amongst the litter of cases and packages.

Addis Ababa must then be the only place in the world that scans all luggage out of the terminal. Why? Well - yes, a very good question but apparently you may be smuggling in a thousand TVs in your suitcase to sell and not pay the duty. There were lots more pushing and shoving to get through the scanning, but eventually we were out into the dark but crowded carpark to meet our Kaleb Hotel shuttle bus – maybe. Sure enough, I walked all over the carpark but could not find the van. I had a number for the Kaleb and tried it, but the phone system was down again - so into a taxi we got. Any chance we were getting a bit "past it"? Surprisingly we were still going okay which says a lot for the resilience we have built up over the last two years, eh?

EE.75: More security Issues

There are huge numbers of men, women, and more particularly teenagers and children in every town and village, and many of them do not have full time employment. Then, put into the melting pot a mix of different tribal groups, different kinship groups and different religions it is probably not surprising that unrest occurs quite regularly. Fortunately, in general, Ethiopians want to be peaceful and happy and just get on with their lives. Injera can keep them alive and very regular bunna (coffee) ceremonies with endless chat means that they are well connected to their communities. Any titbit of information – whether a rumour or factual - spreads like wildfire.

Day 690

Back at the farm, the management have been purchasing Niger seed whilst we have been away for a couple of weeks, and the oil extraction employee had been busy already getting lots of oil processed. It can be a very profitable business for the farm, so Pip and I have put pressure on for the farm to find as much money as possible to buy seed now – whilst it is being harvested by the local farmers. I have also asked head office if they would like to put up some money to purchase seed and we will process it for them under contract. They are hoping to find 200,000 birr (A$10,000) so that would be very helpful. The farm would do well, and head office would double their money according to my calculations. No brainer, you would think – but this is Ethiopia so we shall wait and see.

Moti told us about some demonstrations around here last week whilst we had been away. We had heard that the road past the farm had been blocked and yes, he told us they had not been able to move from the farm for days as the local people had set up roadblocks and barricades everywhere in the district. For the first time anyone could remember, the children and students had been pushed to the back of the demonstrations, and the adults had been standing up to the police and copping the flak. He told us that the main issue for the adults was the way the military had responded to the demonstrations last year and the killing and injuring of their children.

The local farmers had managed to get their cattle, sheep, and goats to the huge Guder market last Monday (it is the biggest livestock market in the western part of Ethiopia) but there had been a disturbance at the market, so the military went in with tear gas!!!! Don't tell anyone with an ounce of compassion for animals (nor actually with compassion for humans) but imagine the consequences.

Given there are no pens, just thousands of men and young boys standing around in incredibly crowded conditions herding their little mobs of animals, and equally large numbers of people pushing their way around trying to find the best deal, one cannot imagine the consequences of a few canisters of tear gas lobbed into the crowd by enthusiastic soldiers!!! Moti seems to think the locals are not impressed. I suspect he is very right.

Late in the day, we walked down to the cow shed at milking time and even the milking ladies had to extract themselves out from under the cows to give us huge welcoming hugs as we had been away for our Christmas break. We were delighted to find that four more cows had calved successfully, so milk production continues to increase. In the heifer shed, we were really pleased to see another eleven heifers pregnant or ready for joining, so the future is surely bright for the dairy if the management can keep up.

We also found that a hectare of chillies had been planted amongst the new sugar cane which is growing like mad. The chillies had fortunately been watered once before the pump broke down.

A busy day, and after a yummy home cooked dinner, it was great to once again head to our comfortable bed in our little hut.

EE.76: Ethiopian Christmas

By now there were very few periods of festivity left before we come to the end of our two-year assignment and pack up and head back to Australia, so it was great to be able to spend Ethiopian Christmas in our community.

Day 692 - Ethiopian Christmas Day

Time to celebrate another Ethiopian Christmas and as we opened our bedroom window to let in the first rays of sunshine, and the lovely fresh air after the overnight rain, we could clearly hear the sounds of excited men's voices, so there was no doubt an ox was being slaughtered outside.

It turned out that the managers and guards had purchased a large ox a few days ago and had slaughtered it this morning and would butcher it during the day. The day labourers had also bought a smaller one and they slaughtered it 100 meters away. A happy day for all and sundry with mountains of raw meat to consume as the day went on and even the local dogs and vultures were happy as they swept in to clean up the remains.

I can handle it okay looking at the meat as it is divided up, and even the carefully divided portions of raw offal. But somehow, I get a bit queasy at the intestines getting chopped up into 3 cm lengths and thrown onto a metal plate warmed up on a fire for a bit of a sizzle and then eagerly consumed by the workers as they butcher up the rest of the carcass. Half-digested oxen poo, half washed out of the intestines which are then half cooked, leaves me a bit cold as a Christmas delicacy.

Cutting up the slaughtered ox and dividing each portion for the families who had contributed to the purchase.

Late afternoon we had a thunderstorm with more rain, and then at 7.00pm we were invited down to Aberash's hut. When we got there, Hordofa was there plus his half-brother and a young female cousin. They had already eaten but Aberash cooked up some of the ox that had been slaughtered today (without spice, especially for us) and we were served it with injera. We continue to reflect on the incredible generosity of people who have so little but are prepared to share food, friendship, love and caring for we foreigners who have so much by comparison.

The power had gone off late in the afternoon, so the entire district was very dark and damp (great for Christmas festivities!!!) and we ate our dinner by candlelight. That sounds rather romantic but eating tough, freshly killed oxen with lashings of injera with your fingers, at a wooden table in a bare mud hut, with growling feral dogs outside the door, and five people with whom you share virtually no common language, is anything but romantic. But it is certainly an experience that is better than any travel (worldly) experience you could ever wish for. And then Aberash sits on her little stool and brews the most beautiful coffee imaginable. Pip only drinks one and I restrict myself to two (the locals always have three) but even then, we know we will be "buzzing" at some stage in the night, thanks to the caffeine.

EE.77: Visiting the British Embassy compound and International Aid

How is it possible that the tax payers of developed nations, along with large and small aid organization across the world, and private individuals, all continue to pour money into Ethiopia decade after decade, but tens of millions of people continue to live in total poverty, and extreme hardship can be found around almost every corner.

Day 702

On this visit to Addis, we were lucky enough to be invited to stay at the British Embassy compound. I could watch the Australians playing cricket against India on the TV and Pip enjoyed tinkling away on the piano. We had a good walk around the compound, including right up to the back corner on the track between the perimeter fence and the jungle, a spot from which you get wonderful views of Addis.

A few months ago, a leopard was spotted (and photographed) on the golf course, and I think it has subsequently been seen on a number of other occasions. Our hostess Hasnija, who works at the embassy and is the wife of the security head for the compound, has often seen an Abyssinian Civet at her front door. It is like a black and white spotty badger, and they are renowned for their musk scent used in perfumes. Chanel No 15 we believe!

During the afternoon, Hasnija, who was born in Bosnia and just survived the siege of Sarajevo as a child, had a friend come to visit who is a fellow Bosnian and works for an NGO in Addis. We spent a lot of the afternoon drinking specially brewed Bosnian coffee and eating delicious sausage rolls, lamingtons, and blueberry buns which Hasnija just whipped up.

Before coming to Ethiopia, I had only vaguely heard the term NGO (Non-Government Organization) but here it is an acronym on the lips of everyone from the Prime Minister to the humble peasant farmer. I have absolutely no doubt that it is the largest sector of the economy and as such causes huge distortions in the economy as the economy attempts to grow.

At the end of virtually every road in Addis there is a sign pointing up the road, notifying of the non-descript office of yet another NGO. Be it the International Fund for Animal Rights, to numerous HIV prevention organizations, to huge multi storey offices of the Save the Children fund. Compounds brimming with new Landcruiser's with World Vision written on the side, or the World Climate Change Institute, Reafforestation projects, Women and Children support and Orphanages. There are thousands, if not tens of thousands, of different organizations. We are all left to speculate as to why, with so much fire power available for help, the country is still in such a mess.

The Bosnian lady works for an NGO called International Rescue Committee which we had never even heard of, but it has one thousand employees in Ethiopia. They spend their efforts helping to provide food, water, and sanitation in rural areas, particularly in the Ethiopian refugee camps which are some of the largest in the world and we are told currently accommodate 900,000 people – asylum seekers into Ethiopia!! Many camps, particularly along the Eritrean and Sudanese borders, have been there for forty years, but the tens of thousands of people in them still have no jobs and rely almost exclusively on NGOs for support. It all seems so hard to get your mind around when you hear of Ethiopian asylum seekers in Australia and elsewhere.

Many of the NGOs are partly funded by the United Nations but the money chain is so long through this source that it makes for hard listening to understand the percentage of the funds that finally get onto the ground. The donor countries first pay their money to the UN and the huge bureaucracy that it requires. The UN then diverts the money to UNESCO and UNICEF – equally large bureaucracies. The money is then sent off by them to a special bureaucratic organization here in Ethiopia which coordinates the distribution of funds to the different NGO aid organizations. Hopefully they then get some of that money onto the ground, along with private donations they have received from people the world over.

The stories we hear make us happy and proud to be working at the coal face and ensuring that the money put up by the Australian people through the Government, finds its way right to the grass roots and that our "in country" support is Demiss who does not have (or even want) and office, just a phone and his car.

Whilst explaining the NGO industry to us, this lady was very strong on the current "emergency" situation caused by the drought which is likely linked to El Nino. She says that donations from around the world are only half of what they need to avoid a human catastrophe with 10.2 million people now facing famine. It sounded to me very much like the Government line which I am a bit cynical about, as we have travelled to almost every corner of Ethiopia in the last two years and have not heard any talk of widespread drought. She argues that the Government does not want to be seen as failing its people and therefore is downplaying the situation "only" listing 10.2 million (the "Live Aid" famine of 1980 only listed 8 million!!!) Some of we cynics take the other view and think the Government is "going big and going early" to ensure it taps into world aid before other countries get a look in.

What we do know beyond doubt is that disastrous farming controls and practices coupled with staggering population growth is a massive contributor to what is described as "the drought." Droughts, or slightly drier than usual times, will always be disastrous if, at all times of every year, all available pasture is grazed into the ground and all animals are on the brink of starvation year-round, because there are simply too many of them.

Without wishing to sound like know-alls, it is this scenario which we recognised from living for two years out in the countryside and having farming experience, but which the NGO officials in the lovely big 4WDs living in the cities simply have no idea about. Suddenly there is a famine and it's all hands-on deck (rightfully) trying to save lives. In actual fact, it is only 10 percent worse than yesterday, but the proverbial camel's back is broken.

It was an interesting afternoon and having eaten platters of delicious food, we only needed a bowl of soup for dinner. We had already had the customary Gin and Tonic – surely essential if you are in a British Embassy compound looking out over gracious buildings wrapped in cloaks of bougainvillea and surrounded by acres of rolling, manicured lawns - and outside the bombproofed, razor wire topped walls - a mere ninety or one hundred million people almost all struggling to survive. What a double life we lead!!!

Chapter 18

Jan 2015 - Goodbye Birbirsa

EE.78: Friendships and "Family"

Now, almost at the end of our assignment and the time had come to cope with the logistics and emotions of our final farewells.

Day 706

This evening as I sit listening to music on our trusty Bose speaker that in Australia had cost us the equivalent of about six months earnings of one of our farm managers here, I am trying to come to grips with a return to the comfortable lifestyle of Australia after only eight more sleeps in our bed at Birbirsa. It is no wonder we have mixed emotions about the end of our assignment, and I am sure there is to be plenty more emotional moments in the next couple of weeks.

After morning coffee, we had one of the most interesting and loving experiences a traveller could have anywhere in the world – albeit in one of the most "financially disadvantaged" places you could find anywhere. Pip has been keen for months to learn how to tie a bag of teff onto a donkey. It is an amazing skill and learnt very early in life by most rural Ethiopian people. Perhaps because today was a "holiday", we had been invited to go the farm of Tyto who had the day off from her cheesemaker duties here at the farm.

Her son, who is about 16 years old, was sent to lead us to her hut. As we walked up the road for about three kilometres, a few others joined the walk, plus a little girl no more than three or four who had taken my hand and had no intention of letting go. As usual I wondered what our responsibilities were as she ventured kilometres from her home territory.

We noticed some large rocks still on the middle of the road in a number of places, presumably from the barricades that were erected across the roads throughout the Oromia region whilst we were away last week. Nobody has got around to shifting the largest of the rocks which would take two or three people to roll off the road.

After about three kilometres, we farewelled our hangers-on, including my little girl and Hordofa, as they kept walking further up the road to Mutulu for the celebration of Timkat, and we headed off across the countryside. Another couple of kilometres and we came to an area that looked like an oasis and was obviously the huts and farmland of the family group. We know that Tyto is a great worker, as is her husband, and this area had their stamp all over it. Lots of stacks of teff being threshed by the men, areas of real bananas and false bananas (ensat), coffee bushes, Khat, vegetables, avocado and mango trees, and a special area of beehives. Their side of the valley is totally different soil type to here at our farm and they can grow all of this without irrigation water even though we are at the same altitude. They do have a potable water tap in the yard which must be quite a long branch line off the system that we are on. They do not have power this far off the road.

Apart from the son who led us to the area, Tyto also has two teenage daughters and twin younger daughters about 7 years old. Also at the hut was a younger sister of Tyto - a girl in her early twenties. She had recently returned from Saudi Arabia where she had been a domestic servant and had suffered abuse. Seeing the distant, troubled look in her beautiful eyes, I have no doubt as to the voracity of that. The Ethiopian Government has recently "forced" the return of something like 100,000 domestic working girls from the Middle Eastern countries due to numerous reports of abuse.

Other members of the families live in the huts in the area, and their children were also about, as was the grandmother. We got the most fabulous, loving welcome from all and sundry.

To describe living conditions as basic – no, very basic – would not be an exaggeration and yet we felt totally comfortable in the surrounds with these delightful, happy, and obviously proud people. We could not help but notice the use of the various containers that we had handed out over the years to staff being put to good use storing sugar, coffee, and spices.

The coffee preparation was left in the hands of one of the daughters whilst Tyto, Pip and Grandma took to the donkey which had been rounded up and brought back to the hut ready for the tying lessons. It was wonderful to stand back and watch the three "girls" chatting and laughing together as ropes wound hither and yon and the very patient donkey had the bag of teff tied and retied over and again. I kept the camera whirring with shots and video to jolt Pip's memory when she is trying to get it right back in Australia.

Back into the hut for delicious bunna, and we did actually drink tella, the home made, roasted and fermented barley beer which is very much a local speciality. Hopefully we do not pay the price of drinking a brew made from the local water – something we do not normally do but this was a very special event with part of our Ethiopian family. We then had injera with scrambled egg – obviously prepared especially for us as there was a large pot of heavily spiced wot put aside which the family were going to eat with their injera when we left.

After lunch we went to visit other areas of the farm including the bees, and more and more family members appeared, and we felt comfortable taking lots of photos as they were proud of what they had to show us. The brother-in-law even came out dressed in his full bee - handling suit, complete with wire mesh head protector. He was obviously enormously proud of it.

Eventually we had to say very fond farewells which including Tyto saying a very heartfelt 'gellatoma' (thankyou in Oromo language) for giving them the honour of coming to their home for lunch. Truthfully, the honour should be completely reversed as what an honour and delight it was for us.

One of Tyto's sisters walked us home, and then she walked on down the road for the seven kilometres to Guder where she works in a café.

Many of the local café/restaurants have a few hotel rooms at the back which are very basic and are mostly rented by the hour by couples who need some privacy which is never available anywhere else!! These rooms cost about A$2 for a night so I suspect an hour or two comes pretty cheap, but at my age if I asked the price for an hour, it may raise a few eyebrows in the town!!!

Day 713

We ticked off a few more final jobs, including moving all my tools onto the shadow board that Pip had made up in storeman Abissa's old shipping container. It seems the most likely place to keep them stored and safe, but the downside is that Abissa then must be found if anyone wants a tool - so quite a compromise.

Lami was sticking very close all day as he watched Pip carefully organizing the distribution of our worldly belongings to the very grateful farm staff.

I asked Moti why the service vehicle headed to Guder again and he muttered some excuses, but not until about lunchtime did we find out that the farm staff had all made a financial contribution to buy some meat and soft drinks and party items to give us a send-off. We were really, really moved that these people had probably committed half a day's pay ($A 60 cents) each to make up a party to farewell us. Moti told me that they wanted to do it themselves rather than the farm to pay as it would try to show us how much they appreciated us and how they would miss us. I cannot imagine anything more humbling.

They put up signs bidding us farewell and giving their blessing to our future and begging us to come back and visit. It also felt great that the ladies came and went from our hut to get things (plates etc) that they needed, and Kidist cooked up the tibbs on our campfire just as if they were in the home of one of their community members – which of course is how we feel.

We had Fanta, popcorn, and lots of injera. I doubt most of the staff enjoyed the food as much as usual, as there was no chilli powder put into the tibbs for our benefit. There was a huge loaf of freshly baked local bread (dobo) which Pip and I had to cut together, with a large knife – memories of our wedding cake 35 years ago.

Only this morning Pip had said how it looked like we would not get a chance to show the photo/video presentations that she had made up for the staff a few weeks ago, and for which our son Richard had brought us over a projector to connect to our computer. Suddenly we were presented with the perfect opportunity to set it up and run the videos (so often it all does come together at the last minute in this country). All afternoon we had the photos rolling on a big screen (a white, biogas bag). They were mostly of the staff members, children and activities around the farm and brought much delight to all concerned.

Tibebu and another couple of senior community people also came, and all the managers and guards and some of the labourers kept coming and going.

It was quite a magical afternoon, particularly when the oldest staff member Katila who is a farm guard was asked to bless us. He went on for a few minutes with a special chant, and all the people present responded with chants at different times. Tibebu was then asked to translate, and he relayed that Katila had been saying lots of lovely things about us and wishing us well for the future. Lami sat on my knee pretty much throughout which always amuses the adults as they rarely have physical contact with children – at least outside of their homes.

Late in the afternoon everyone drifted off, and we put everything away. We handed out a bit more stuff so there is just enough for us to get through with noodles and scrambled eggs for dinner tonight, our bed to sleep in and enough for breakfast. The power went off just on dusk, and some light rain fell – an inevitable farewell. Lami and his mother Kidist were around our hut until well after dark.

Day 714

By now we are well versed in the Ethiopian clock, so we told Moti that we wanted to be on the road by 4 am Ethiopian time which is 10am by our clocks. Moti has never wavered from using the Ethiopia clock. As this load to Addis in the old Hilux farm ute was going to include all our personal belongings, I was determined that it would be loaded to my requirements, and I would tie down the load.

Into the ute went four hundred and fifty litres of Niger seed oil, one hundred and twenty kilograms of cheese and one hundred kilograms of butter. Back up to our hut where we loaded on our double bed and mattress and one of our big office chairs. All of which we were giving to Demiss to thank him for all the outstanding care and friendship he had given to us for the last two years. He is building a new house in Addis, so it was very timely for him and his family.

Then I loaded our suit cases and other bags on top and roped it all down well. It all seemed so unpleasantly final. As we were doing our last packing, Pip counted the crowd at thirty with lots of children right under our feet every time we moved.

Some of the assembled gathering for our farewell.

Just before ten we went into our bedroom to change into our Addis clothes, (having had to hunt out six kids who followed us in), and then went and distributed our last (dirty) shirts, jeans, and work boots.

Our two suitcases were almost empty having left as much as we possibly could with these needy people, but our hearts were completely full of love and admiration for this community and country.

Finally, numerous hugs and kiss from everyone and lots and lots of damp eyes. By now Lami had disappeared as we expected, as he is not into farewells. He was well aware that we would not be back in the short term, but he told me earlier that when we come to visit, he will be a big boy and going to school.

Emotionally and physically, we were ready to return to life in Australia but equally not at all ready to leave this unique country and wonderful Birbirsa community.

Departed but never Forgotten

To close the final chapter of our Ethiopian Experience, by chance I sit writing in a hotel room in Addis Ababa, Ethiopia some years after our volunteer assignment. Pip is always happy to be looking after our farm in Australia – as she says – well away from the chaos that IS Ethiopia.

My love for the country and its people, and our "Ethiopian family" goes deeper, and I have returned many times. I have taken the opportunity to take several small groups of Australians to Ethiopia and guide them through the back blocks of the country.

On every occasion, I spend some time at the farm in Birbirsa and continue in our farm management mentoring role. Two years after our assignment, the farm was transferred from AGOHELMA (now called Abebech Gobena Charity) to the university of Ambo. Pip and I are really proud that it is now very productive, and the focus of the farm is very much on the direction that we pushed – particularly towards the dairy and pasture management. Hopefully students from the agricultural faculty of the university are learning and observing the improved farming techniques and management and, as they graduate, they are taking that knowledge back to the agricultural sector throughout Ethiopia.

Through the generosity of the Australian people and resources of the Government being directed into the Australian Volunteer program, Pip and I are confident that Australia has made a significant contribution to the productivity of the agricultural sector in Ethiopia through our assignment.

Our lives will forever be enriched by the experiences – the need to develop a strong moral compass, an empathy with extreme poverty and hardship, a tolerance for perceived inefficiencies created by centuries of custom and tradition and becoming totally immersed in the culture where we lived for two unforgettable years.

More than eighty percent of Ethiopians live in the countryside. It is there on farms, in scattered villages, and large towns that Ethiopia truly reveals itself. You can trek through the highlands of the Simien Mountains, marvel at the amazing national parks on the dry flat areas in the east of the country, lose yourself in chaotic downtown Addis Ababa, or stand in awe before the world-famous rock-hewn churches of Lalibela in the north and the ancient cultures of the Omo Valley in the south. All of these are extraordinary. But they are only the surface.

The real beauty of Ethiopia does not lie in its monuments or its landscapes—it lives in the hearts of its people as two very ordinary Australians were to discover.

Glossary of Ethiopian Terms

Araki – A strong traditional distilled alcoholic spirit, usually home-produced and widely consumed in rural areas.

Bunna – Traditional Ethiopian coffee, prepared and served in a formal coffee ceremony.

Dobo – Bread made from enset (false banana), typically from fermented and processed enset flour.

Enset – The "false banana," a staple food plant in southern and highland parts of Ethiopia. The pseudo stem and root are fermented, dried, and processed into flour for bread making.

Farangi – A common expression used to describe foreigners, particularly Europeans or Westerners.

Injera – The staple food of Ethiopia: a large, round, slightly sour fermented flatbread made from teff flour and eaten with almost every meal.

Kitfo – Finely minced raw (or lightly warmed) beef mixed with chili and spiced butter, traditionally served with injera.

Shiro – A thick stew (wot) made from ground chickpeas or broad beans, usually spiced and served with injera.

Tibbs – Fried or sautéed pieces of meat, usually cooked with onion, butter, garlic, rosemary, and chili.

Tella (Tulla) – Traditional local Ethiopian beer, home-brewed from grains and fermented.

Wot – meat or vegetable stew

Key People

Key People Involved with AGOHELMA and the Birbirsa Farm
AGOHELMA Organization Leadership

Abebech Gobena

Abebech Gobena was the founder and spiritual leader of the AGOHELMA organization. She graciously welcomed us into the AGOHELMA family and maintained a close interest in our activities throughout our time there. She lived in the Addis Ababa compound of the organization, and we mostly paid her a visit when we were in the head office. Sadly, by the end of our assignment, her health was deteriorating, and she passed away in 2021 due to Covid.

Ato (Mr) Eshetu and **Ato Tesfaye**

Serving as the general managers of AGOHELMA, Ato Eshetu and Ato Tesfaye were based at the head office in Addis Ababa. They played a central role in overseeing the operations and administration of the organization.

Farm Personnel

Tibebu – Farm Manager (our First Year)

Tibebu served as the farm manager during the first year of our stay at Birbirsa. He is a born leader and a charming, intelligent man with great experience and a wonderful sense of humour. He had been manager of the farm since its inception. His presence at the farm was not always consistent, which meant that other staff often took on greater responsibilities in managing day-to-day activities.

Lechisa Angussa – Birbirsa Farm Overseer (our First Year)

After earning a degree in Agricultural Science from Jimma University, Lechisa returned to his home village of Birbirsa and began working as the farm overseer at the AGOHELMA farm. Prior to our arrival, he lived with his wife, Zinash, in Guder and walked seven kilometres each day to the farm, while Zinash was still attending school in Guder. Because the farm manager, Tibebu, was frequently absent, Lechisa was often responsible for the daily operations. We formed a close relationship with him, and when Zinash completed her schooling, the couple moved into a room near us at the farm. This arrangement brought us great personal joy and allowed us to help develop Lechisa's management skills and agricultural expertise. Although we hoped Lechisa would be promoted to farm manager, his growing reputation led to him being recruited to manage a large coffee plantation in southwest Ethiopia, where he now oversees hundreds of hectares of coffee plantings and numerous staff members.

Moti Debele – Birbirsa Farm Overseer and Manager (our Second Year)

Moti had a background in agricultural business management and forestry and was eager to expand his knowledge of agriculture and farm management. We worked closely with him, and his appointment brought valuable business skills to the farm when they were much needed. After Tibebu's retirement, Moti was promoted to farm manager. A year or two after we left the farm, Moti left to become a university lecturer and was succeeded by two young graduates in agricultural science.

Jagoma – Foreman of Works

Jagoma, appreciated by Lechisa for his skill in organizing the day labourers, played a key role in managing the workforce. However, he was not favoured by Moti, and as a result, his position was eventually eliminated.

Kidist (and Son Lami) – Farm Cashier

Kidist, along with her two-year-old son Lami, were the only permanent residents on the farm when we arrived. Living in a hut about 200 meters from us, they became very close friends—Lami became like a grandchild to us, spending most of his days with either Pip or me. As the farm cashier, Kidist collected cash payments, deposited them in the bank, and arranged for cash distributions via the purchasing officer and the head office in Addis Ababa, including monthly staff wages. Pip spent many hours teaching Kidist computer skills, and her confidence in English grew significantly. Lami also quickly became fluent in English through daily interactions with Pip.

Aberash (and **Hordofa**) – Farm Shopkeeper

Aberash lived near the farm with her twelve-year-old son, Hordofa. She managed the sale of farm products—including milk, butter, cheese, Niger seed cooking oil, and seasonal vegetables—from a building on the farm and assisted with sales in nearby towns and villages. Hordofa, who visited the farm daily after school, was intelligent, artistic, musical, and already had a good command of English. Pip taught him touch typing and computer skills, and we trusted him completely. He often served as our interpreter and was a vital bridge to the local community. We have since supported his pursuit of a medical degree.

Tyto – Cheese and Butter Maker

Tyto was a dedicated and industrious woman from a neighbouring farm who worked part-time processing milk into cheese and butter, and occasionally relieved as a milker. She always treated us as if we were part of her family.

Merka – Farm Driver (our First Year)

In our first year, Merka was the farm's driver and the only person qualified to drive both the Hilux ute (the service vehicle) and the tractor. After Pip and I obtained Ethiopian drivers' licenses, we were able to drive the vehicles ourselves and teach the dairy manager, Getun, to drive. Merka then left the farm for a job as a government driver in his hometown of Ambo.

Getun – Dairy Manager, Later Driver

Our confidence in Getun as a dairy manager and a driver was initially low, as he had a license but no real driving experience. Through Pip's patient instruction in the Hilux ute, and my guidance on tractor operation, he improved significantly. He continued as dairy manager under close supervision and ultimately became a full-time driver. He now credits Pip with transforming his life and that of his family with security and a good income.

Abissa – Manager of Stores

We affectionately called Abissa "Mr. Keys" because he was always seen with a large collection of keys. He was responsible for ensuring that all purchases were delivered and stored securely, and that all sales were properly recorded as the items left storage. Locating him to access tools or supplies was sometimes challenging, as he often spent time at his nearby home and farm with his large family. However, we enjoyed the company of his delightful children, who frequently visited our hut.

Abara (and **Barjar**) – Guard

Among the six farm guards, Abara held a special place in our hearts. We became acquainted with most of his eight sons, with Barjar becoming particularly close to us and Lami, spending much time together as friends. Abara appreciated our support for his family and was always ready to help, despite the language barrier. He is one of those people who can sense when you are struggling and would often light a campfire or collect water for us to give us not just physical support but also emotional support. Simple kindness. Along with the other guards, he was dedicated to our safety and well-being.

Other Guards: **Katilla**, **Carson**, **Lechisa**, **Angussa**, **Camilla**

These men were steadfast in protecting the farm and caring for Pip and me. At any given time, at least two guards were on duty - one near our hut and another monitoring the eucalyptus plantations and cattle sheds.

Other Farm Workers

Bikele (milk processing), **Abara** (cowboy), **Abebech** (chickens), **Worku** and **Kitinish** (day workers), and **Zinash** (shepherd), and the milking ladies (whose names I never did manage to pronounce correctly – let alone spell) were all essential contributors to the daily operation of the farm. We worked alongside them with great affection and mutual understanding.

Support in Addis Ababa

Stuart (and Linda) Shaw – Consul General to Ethiopia and Surrounding Countries

Stuart and his wife Linda became dear friends, and we often stayed at their house in Addis Ababa. Numerous times we arrived in our old Toyota Landcruiser, tired and unwashed and with loads of washing, and spent a day or two of rest and recreation in their beautiful three-story house. On longer stays in Addis Ababa during our second year we used the house as a base as we attended conferences, meetings or researched projects for the farm at Birbirsa.

Alan (and Hasijna) Lee – Senior Security Advisor to the British Embassy in Addis Ababa

Introduced to us by Stuart and Linda Shaw, Alan and his wife, Hasijna, welcomed us warmly. We became regular visitors to the British Embassy, enjoying many pleasant social times with the Lees.

Mark (and Anaise) Sawers – Australian Ambassador to Ethiopia and Surrounding Countries

Mark and his wife, Anaise, were approachable and friendly, hosting us for numerous embassy formal and social events. On several occasions, we stayed overnight in the guest’s room at the ambassador's residence.

Made in United States
Troutdale, OR
11/03/2026

56844379R00166